Purposeful Engagement in Science Learning

This book is part of the Peter Lang Education list.
Every volume is peer reviewed and meets
the highest quality standards for content and production.

PETER LANG
New York • Bern • Frankfurt • Berlin
Brussels • Vienna • Oxford • Warsaw

Kabba E. Colley

Purposeful Engagement in Science Learning

The Project-based Approach

PETER LANG
New York • Bern • Frankfurt • Berlin
Brussels • Vienna • Oxford • Warsaw

Library of Congress Cataloging-in-Publication Data
Names: Colley, Kabba E., author.
Title: Purposeful engagement in science learning:
the project-based approach / Kabba E. Colley.
Description: New York: Peter Lang, 2016.
Includes bibliographical references and index.
Identifiers: LCCN 2016005904 | ISBN 978-1-4331-3091-5 (hardcover: alk. paper)
ISBN 978-1-4331-3090-8 (paperback: alk. paper) | ISBN 978-1-4539-1847-0 (ebook pdf)
ISBN 978-1-4331-3565-1 (epub) | ISBN 978-1-4331-3566-8 (mobi)
Subjects: LCSH: Science—Study and teaching. | Project method in teaching.
Classification: LCC Q181 .C537 2016 | DDC 507.1—dc23
LC record available at https://lccn.loc.gov/2016005904

Bibliographic information published by **Die Deutsche Nationalbibliothek**.
Die Deutsche Nationalbibliothek lists this publication in the "Deutsche
Nationalbibliografie"; detailed bibliographic data are available
on the Internet at http://dnb.d-nb.de/.

© 2016 Peter Lang Publishing, Inc., New York
29 Broadway, 18th floor, New York, NY 10006
www.peterlang.com

All rights reserved.
Reprint or reproduction, even partially, in all forms such as microfilm,
xerography, microfiche, microcard, and offset strictly prohibited.

CONTENTS

	Acknowledgments	xi
	Preface	xiii
Chapter 1.	Introduction to Project-based Science Instruction	1
	Chapter Overview	1
	Historical Foundations of the Project-based Approach	2
	Philosophical Foundations of the Project-based Approach	5
	Psychological Foundations of the Project-based Approach	8
	Project-based Science Instruction in the United States	11
	What Is Project-based Science Instruction?	15
	Categories of Projects	18
	Science Learning Standards and Project-based Science Instruction	19
	Chapter Summary	22
	Food for Thought	23
Chapter 2.	What Does Research Say About Project-based Science Instruction?	25
	Chapter Overview	25
	Conducting the Review	25
	What Does the Research Says about PBSI?	28

	PBSI Curriculum Materials	28
	Engagement of Teachers in PBSI Professional Development	41
	Engagement of Students in PBSI	42
	Impact on Student Learning	43
	Applications of PBSI Within and Across Science Disciplines	44
	Biology	44
	Chemistry	46
	Physics	48
	Earth Science/Environmental Science	49
	Engineering	50
	Chapter Summary	51
	Food for Thought	52
Chapter 3.	How to Plan for Project-based Science Instruction	53
	Chapter Overview	53
	Underlying Principles and Concepts of PBSI	54
	A Day in the Life of a Project	55
	Orientation to PBSI	56
	Expectations and Requirements of PBSI	57
	Roles and Responsibilities	58
	How to Identify and Define a Project	59
	Aligning Students' Project Questions to Science Learning Standards	64
	Developing a Project Plan	68
	Chapter Summary	69
	Food for Thought	71
Chapter 4.	Implementing Project-based Science Instruction	73
	Chapter Overview	73
	What Factors Should We Consider Before Implementing PBSI?	73
	Type of Principal	74
	Teacher Knowledge, Skill, and Disposition	74
	Type of Curriculum	75
	Availability of Funding	76
	Parental Involvement	77
	Level of Technology Adaptation	77
	Physical Environment	78
	Time	78
	Strategies for Implementing Project-based Science Instruction	79
	Teacher-centered	79

CONTENTS

Student-centered	79
Teacher-Student Partnership	79
One Teacher-One Classroom	80
Multiple Teachers-Multiple Classrooms	80
Teacher-Student-Scientist Partnership	80
Extracurricular Activity (PBSI Clubs/Societies and Afterschool Programs)	81
Internships	82
Using Microcomputer-based Laboratories Tools	82
Collecting, Analyzing, and Interpreting Project-based Data	84
Preparing and Presenting Project Reports	89
Chapter Summary	91
Food for Thought	92
Chapter 5. Case Histories of Project-based Science Instruction	**93**
Chapter Overview	93
Case 1. Project-based Biology: The Herpetology Project	93
Background and Context	93
Teacher's Role	94
Students' Role	95
Strategies for Implementation	97
Project Activities	97
Challenges and Lesson Learned	98
Case 2. Project-based Chemistry: Investigating Carbon Dioxide and Indoor Air Quality	98
Background and Context	98
Carbon Dioxide and Indoor Air Quality Project	100
Teacher's Role	101
Students' Role	101
Strategies for Implementation	102
Project Activities	102
Challenges and Lesson Learned	102
Case 3. Project-based Earth/Environmental Science: Park Ecology	103
Background and Context	103
Teacher's Role	103
Students' Role	104
Strategies for Implementation	105
Project Activities	105

	Challenges and Lesson Learned	106
	Case 4. Project-based Physics: High-Altitude Ballooning	107
	Background and Context	107
	Teacher's Role	109
	Students' Role	109
	Strategies for Implementation	109
	Project Activities	109
	Challenges and Lesson Learned	110
	Chapter Summary	110
	Food for Thought	111
Chapter 6.	Evaluating Project-based Science Instruction	113
	Chapter Overview	113
	Methods Currently used to Assess Students in Project-based Science Learning Environments	113
	Assessment by Presentation and Product	118
	Oral Presentation	118
	Strengths of Oral Presentation	121
	Weaknesses of Oral Presentation	122
	Essay	124
	Strengths of Essay Assessment	127
	Weaknesses of Essay Assessment	128
	Project Reports	128
	Drawings	129
	Strengths and Weaknesses of Assessment Using Project Report and Drawings	130
	Portfolios	130
	Strengths of Portfolio Assessment	133
	Weaknesses of Portfolio Assessment	134
	Assessment by Performance and Observation	135
	Strengths of Assessment by Performance	140
	Weaknesses of Assessment by Performance	141
	Chapter Summary	143
	Food for Thought	143
Chapter 7.	Project-based Science Instruction, Afterschool Science Programs, and Community Engagement	145
	Chapter Overview	145
	Purpose of Afterschool Programs	146
	Characteristics of Afterschool Programs	146

	Relationships Between PBSI and Afterschool Science Programs	147
	Community Engagement in Afterschool PBSI	148
	Potential Benefits and Challenges of Afterschool PBSI Programs	150
	Chapter Summary	151
	Food for Thought	152
Chapter 8.	Resources for Project-based Science Instruction	153
	Chapter Overview	153
	Using Library Resources	153
	Using Internet or Online Resources	155
	Museums and Science Education Centers	156
	Computer Hardware and Software	157
	Scientific Tools and Technology	164
	Curriculum and Instructional Materials	168
	Scientific Electronic Databases	171
	Government Agencies	172
	Professional Science and Science Education Associations	173
	Scientific Research Centers and Organizations	174
	Chapter Summary	174
	Food for Thought	174
	References	177
	Index	195

ACKNOWLEDGMENTS

This book was inspired by my students, who just wanted to know how to conduct PBSI without having to take specialized courses or engage in endless professional development activities. To these students, I owe a debt of gratitude for their tough questions and insistence that I provide an example by not only talking the talk but also walking the walk. This book would not have been possible without Dr. Binta M. Colley, my wife and partner in crime, who was never bored listening to my "Colley Theories," who was never too tired to read my drafts at all times of the night and day, and who unceasingly edited my drafts and provided constructive feedback throughout. Thank you my Dear for all your loving support and encouragement. Although it would be impossble to name them all here, I wish to thank those who came before me, on whose shoulders I stand. Finally, special thanks to Dr. Robert F. Tinker, former Chief Scientific Officer at TERC Inc., and President Emeritus of Concord Consortium for his support and mentoring during the embryonic stage of my PBSI journey, and all my former colleagues at TERC, Inc., where the seeds for my interest in PBSI were sown.

PREFACE

The history of projects is as old as the history of human society. Wherever humans have lived, projects have existed. Humans are constantly engaged in activities such as building homes and shelters, constructing roads, and inventing transportation systems. In addition, humans are constantly engaged in farming and food production, and exploitation of the earth's natural resources. All of these activities are purposeful acts, which is what projects are. Projects are defined as "purposeful acts that are conducted within a social context."

Over the last century, the teaching and learning of science from K–16 has been implemented mostly through the use of the lab and lecture method. Although this approach has served the science education community well, it could be argued that this approach has outlived its usefulness. In order to get students excited about science, increase their participation in science and engineering careers, and promote scientific literacy among ordinary citizens, a new approach is required. When we look at the scientific enterprise and how science is conducted in the real world, we find that science, or the practice of science, is dominated by purposeful activities or projects. Science, as is practiced in the real world, is characterized by complex social activities and relationships that involve scientists engaged in a cycle of knowledge theorizing, construction, testing, validation, and sharing.

To encourage students to learn science and to promote scientific literacy, the way science is conducted in K–16 must change to reflect the way science is conducted in the real world. This is no small task and will require a change in not only policy and practice, but also a change in our current mind-set about teaching and learning of science.

This book, *Purposeful Engagement in Science Learning: The Project-based Approach*, provides a blueprint on how teachers and their students can engage in purposeful science learning that mirrors the way science is practiced. This book is written for K–16 science educators as well as those in the informal science education sector. The framework for this book is based on the project cycle, which makes this book flexible and easily adaptable to a wide variety of science learning environments.

The book is divided into eight chapters. Chapter 1 provides background and context to project-based science instruction (PBSI) by reviewing the historical, philosophical, and psychological foundations of PBSI. This chapter also traces the evolution of PBSI in the American science education scene. Chapter 2 attempts to answer the question: What does the research say about PBSI? Chapter 3 looks at how to plan PBSI and offers practical strategies for teachers and students who are veterans and novices to this approach. There are different strategies to implementing PBSI. Chapter 4 takes the reader through these strategies with particular emphasis on factors to consider and roles and responsibilities of teachers and students. Chapter 5 provides selected case histories that have been successfully implemented and are considered examples of what is possible in PBSI. Chapter 6 deals with the different methods of evaluating and assessing students' learning in PBSI. This chapter also offers concrete examples of assessment instruments that could be used to evaluate students' learning in PBSI environments. Chapter 7 examines the relationship between PBSI, afterschool programs, and community involvement. Finally, Chapter 8 identifies and describes relevant resources currently available that could be used to support and enhance PBSI.

This book follows the path of half a dozen or so books that have been specifically written on PBSI in the last two decades. The author owes a debt of gratitude to those pioneers who took on a subject that is not well understood and tried to make it accessible to all those who want to implement PBSI. The contribution of this book to the field of science pedagogy is its specific focus on the project cycle and its attempt to reach diverse audiences both in the formal and informal science sector. In addition, this book is organized in a way that allows science educators to address the Next Generation Science Standards (NGSS), while at the same time helping students learn science in ways that are relevant to their lives.

Kabba E. Colley

· 1 ·

INTRODUCTION TO PROJECT-BASED SCIENCE INSTRUCTION

Chapter Overview

Project-based science instruction (PBSI) did not originate in a vacuum. Rather, it grew out of a particular historical and philosophical context. To situate this pedagogical approach in context, we will begin this chapter by examining its historical, philosophical, and psychological foundations. The chapter will then trace the history of PBSI in the United States. Various authors, researchers, and science educators have defined project-based science instruction. However, these definitions vary by context and discipline. This chapter will also examine the various definitions of PBSI with a view to providing a comprehensive working definition. The relationship between science-learning standards, nature of science, and PBSI is not well understood and warrants further discussion. This chapter will end by reflecting on this relationship with a particular emphasis on the Next Generation Science Standards (NGSS) and PBSI.

Historical Foundations of the Project-based Approach

> We note, first of all that a very large proportion of the fruitful experiences of life are the purposeful type. We make permanent changes in our behavior while we are in the pursuit of ends. Consequently we must provide a large number of opportunities for the most educative kinds of purposeful activities in school. Attempting to do this affects in turn our notions of what children should be learning, of the kind of teacher we want, the kind of course of study he [she] ought to have in hand, how this should be made, the choice and use of textbooks, specifications for buildings and equipment, methods of supervision, testing, and promotion, as well as our relations with the parents—to say nothing of the janitor. Aims, subject matter, activities of pupils and teachers, standards of achievement, measures of results—all of these are passed in review, and all are evaluated afresh in the light of this far-reaching conception of life and learning.
>
> —Hosic & Chase (1924, p. 6)

Historically, all societies, regardless of time and place, have used projects in one form or another to educate their young and continue to do so up to the present moment. In predominantly agrarian societies, the majority of the people worked on the land. They conducted activities such as preparation of land, cultivation of crops, rearing of livestock, fishing, and hunting. Much of that changed with the advent of the Industrial Revolution. In modern industrial societies, most of the population is employed in the processing of raw materials, manufacturing of goods, and provision of services. These activities that people engage in, whether in agrarian, modern industrial, or societies in transition (those societies that are not purely agrarian or industrial), are project-based because they are purposeful activities punctuated by clear starting and finishing points and conducted within specific social contexts (Kilpatrick, 1918).

In most societies education is structured to mirror and reproduce that society. Project-based instruction is an attempt to teach children or learners by having them engage in worthwhile activities that mirror the societies they live in. As a result, the history of the project-based approach cannot be separated from the history of human society. It is beyond the scope of this book to provide a detailed historical account of education in the United States. However, a brief review or survey will help put this issue in perspective.

One of the educational historians who captured the educational context in early America so well is E. H. Gwynne-Thomas. In his book *A Concise*

History of Education to 1900 AD, Gwynne-Thomas (1981) presents a vivid account of education in Western Europe and America. According to the author, during the founding of the American Republic, there was no mention or discussion of public education. The idea of education for the masses was left to the states to decide. As a result, by the 1780s, most of the original thirteen colonies had included in their constitutions a particular position on public education. One state in particular, the Commonwealth of Massachusetts, included a very bold statement about education and schooling in its constitution of 1780 as follows:

> Wisdom and knowledge, as well as virtue diffused generally among the body of the people, being necessary for the preservation of their rights and liberties, and as these depend on spreading the opportunities and advantages of education in the various parts of the country, and among the different orders of the people, it shall be the duty of the legislatures and magistrates, in all future periods of the Commonwealth, to cherish the interests of literature and the sciences, and all seminaries of them. (Gwynne-Thomas, p. 192)

However, despite constitutional statements about public education, most states did not provide a comprehensive public education system that was accessible to all citizens until after the Civil War, and subsequently, after the passing of the *Brown v. Board of Education of Topeka*. Gwynne-Thomas goes on to trace the evolution of state colleges and universities, and notes that the University of Georgia was the first to be chartered, but the University of North Carolina was the first open. New York State attempted to establish a state university system based on the French model and New Hampshire tried to incorporate Dartmouth College into a state university system but was unsuccessful, because if that were to go through, it would mean that the charter of Dartmouth would be eliminated. Thomas also notes that in 1785, shortly after the American Republic was founded with the Treaty of Paris in 1783, an ordinance was passed to survey the Northwest Territories, which, although not part of the original thirteen colonies, were handed over to the United States (although Native Americans will disagree with that version of history). The ordinance required that the "sixteen section of each township be reserved for the support of education" (Gwynne-Thomas, p. 192).

Bowles and Gintis (1976) maintain that the expansion of the industrial capitalist economic system in America was one of the major drivers of educational reform and expansion in the Antebellum Period. Prior to this time, the responsibility for educating the child was left to the family, apprenticeship activities, or the church. According to the authors:

> The child learned the concrete skills and adapted to the social relations of production within the family. To put the point more technically: Production and reproduction were unified in a single institution—the family. Preparation for life in the larger community was facilitated by the child's experience with the family. ... It was not required that children learn a complex set of political principles or ideologies, as political participation was limited. The only major cultural institution outside the family was the church, which sought to inculcate the accepted spiritual values and attitudes. In addition, a small number of children learned craft skills outside the family as apprentices. Elementary schools focused on literacy training to facilitate a familiarity with the scriptures. Above this level, education tended to be narrowly vocational, restricted to preparation of children for a career in the church, the "learned professions," or the still inconsequential state bureaucracy. (p. 156)

The establishment of the early elementary schools was an extension of these informal structures of education. To support their claim, Bowles and Gintis reviewed statistical records and available primary historical sources on economic life and schooling in New England—where some of the earliest public elementary schools were established—between the 1820s and 1860s. In addition, they conducted a case study of the manufacturing town Lowell, Massachusetts, and the major education reformer at that time, Horace Mann, Secretary of the Massachusetts Board of Education. Their analysis showed that there was a strong relationship between economic development, demographics, and mass public education. The authors cite examples of class struggles between the mill owners, wealthy artisans, and shopkeepers (business and professional classes) on one hand, and the "shoemakers, farmers, sailors and laborers" (p. 155) (working classes) on the other hand. For instance, in 1860, during a town meeting in Beverly, the working classes voted unanimously to close down the town's newly built school over a high school tax and labor-related grievances with mill owners. However, a coalition of wealthy industrialists-capitalists and education reformers were able to exert enough power and influence to establish mass public education throughout New England because, as Bowles and Gintis put it, "Educated workers, they noted, would be better workers" (p. 161).

Expansion of mass public education in the South and West of the country was a different story. In the South, plantation and the slave-owning classes resisted the spread of education because of their racist attitudes toward African/Black slaves and because they saw educated slaves as a threat to the system of slavery. The post–slavery and Reconstruction era led to the expansion of mass public education in the south, thanks to the activities and investments of Northern capitalists, who saw an educated wage labor force as vital to the growth and modernization of agricultural production in the South. By the late 1800s and early 1900s, with the help of their educational philanthropic organizations such as the Ford Foundation, John D. Rockefeller, Peabody Fund, American Bankers Association, and some railroad companies, to name a few, these capitalists used their money and influence to support agricultural education and extension education programs throughout the southern and western states. This era also began the "separate and unequal" public education system in America for decades until the landmark ruling by the United States Supreme Court in *Brown v. Board of Education of Topeka* in 1954. The ruling overturned the *Plessy v. Ferguson* decision of 1898, which sanctioned state-supported segregation of public education in America. In the West, Populists were prominent in challenging the expansion of mass public education. According to Bowles and Gintis, although Populists supported the idea of public education, they saw the country's priorities differently. They believed that the problem facing America at that time was economic inequalities and poverty, and it was more important to focus on changing the structure of the economic system rather than focusing on the expansion of mass public education. With the collapse of the Populist movement, the capitalists and their allies in the ruling and middle classes were able to succeed in their goal of expanding mass public education throughout the country.

Philosophical Foundations of the Project-based Approach

According to Stockton (1920), "Seventeenth-century American education was but a duplication of that prevailing in Europe" (p. 27). This means that ideas that led to the project-based instruction movement in the United States, to a large extent, can be traced to the work of European philosophers-educators like Jean-Jacques Rousseau, Johann Heinrich Pestalozzi, Johann Friedrich Herbart, and Friedrich Froebel (Stockton). Rousseau espoused the theory of a child-centered education in his work titled *Emile*, written in 1762. He proposed that the child should learn through interaction with things and objects. The purpose of education is to prepare the child to live in harmony with nature. He proposed that the cognitive and developmental stage of a child be the determinant of what to teach that child. Although Rousseau's theory of child-centered education was perceived as quite revolutionary, he, like philosophers of his time, did not feel that girls should be educated the same as boys (Noddings, 1995).

Pestalozzi actually applied Rousseau's principles of child-centered education by opening a school for poor children where he combined compassion and learning by doing. He believed that education should aim for the development of the whole child and, unlike Rousseau, Pestalozzi emphasized child-centered learning that integrated knowing and doing within the context of a specific subject matter. According to Gutek (2001), "A strong methodological consistency ran through Pestalozzi's general and specific methods. Both cognitive and affective growth were stimulated by children's sensations. Learning began with objects found in the immediate environment. Progressing slowly, gradually, and cumulatively, without haste or force" (p. 142). During his time, Pestalozzi initiated various reforms in teaching in his native country of Switzerland that became very popular throughout Europe. For instance,

> The conventional approach to instruction was to have all children in one room and to have one child at a time come before the teacher to recite a previously assigned lesson. Pestalozzi believed that this inefficient use of time failed to promote socialization. Instead, he developed the group process of simultaneous instruction to replace the individual recitation. (Gutek, p. 142)

The influence of Pestalozzi was felt not only in Europe but also in America. Several prominent American educators visited him and returned home to implement his ideas in one form or the other. This was particular true for the early progressive educators. Gutek reminds us,

> The spirit of Pestalozzi also could be found inspiring U.S. progressive educators in the late nineteenth and twentieth centuries. These progressives struggled against the continuing residues of formalism and verbalism in many schools. Although, only one of the many influences on progressivism, Pestalozzi's general method sparked the American progressive's emphasis on children's interests and needs. (p. 145)

Herbart was a disciple of Pestalozzi and is credited for extended and formalizing Pestalozzi's ideas in what became known as the *Herbartian teaching method*. This method of teaching could be considered one of the cornerstones of today's structured lesson plans. It proposed five stages in the process of teaching: preparation, presentation, association, generalization, and application (Johnson, Dupuis, Musial, Hall, & Gollnick, 1999).

Froebel, who was also a student of Pestalozzi, put the finishing touches on child-centered education by proposing that the school be a macrocosm of the society in which the child lived. He advocated learning by doing, creativity, and overall development of the child. Froebel believed that by nature every child is creative and these creative impulses needed to be nurtured and allowed to grow through activity and play. He advocated and encouraged children in his care to engage in various activities such as building "dams, windmills, fortresses, and castles, and search the woods for animals, birds, insects and flowers" (Stockton, p. 19).

The idea of a child-centered education, along with other pedagogical approaches, gained fertile ground in the progressive education movement. The progressive education movement, which began in the late 1800s and lasted up to the mid-1950s, was rooted in the philosophy of pragmatism. Developed by Charles Peirce (1839–1914) and extended by William James (1842–1910), the philosophy of pragmatism holds that the meaning and importance of knowledge is found in its practical applications. Pragmatists believed that the best type of learning was the one grounded in experience and that truth comes from knowledge that has been tested and proven to work. The main assumption of the philosophy of pragmatism is that practical experience takes precedence over theoretical knowledge and that it will always be possible to test or apply knowledge. However, one could argue that not all knowledge lends itself to practical application, although knowledge could be critically analyzed and understood. In addition, practical application, devoid of theoretical guidance, is empty since the two are mutually inclusive and must go hand in hand.

The progressive education movement, led by John Dewey (1857–1952), built on the ideas of the pragmatists, and in the mid-1920s established themselves as a major force in U.S. education. They rejected the authoritarian, subject matter–driven, teacher-centered curriculum and advocated for a curriculum that was flexible and placed the child at the center of the learning process. They favored a curriculum that promoted "the scientific method of teaching and learning, allows for the beliefs of individuals and stresses programs of student involvement that help students learn how to think" (Johnson et al., p. 396).

According to Stockton, "If the work of any one man [or woman] were taken to represent the modern situation with regards to actual problems of instruction as a whole, the best one for our purposes would be John Dewey" (pp. 38–39). As an educational philosopher, Dewey (1916) argues for the social function of education in a democracy. He notes:

> Democracy cannot flourish where the chief influences in selecting subject matter of instruction are utilitarian ends narrowly conceived for the masses, and, for the higher education of the few, the traditions of a specialized cultivated class. ... A curriculum which acknowledges the social responsibility of education must present situations

where problems are relevant to the problems of living together, and where observation and information are calculated to develop social insight and interest. (p. 192)

Although the above statement does not explicitly mention project-based instruction, inferences can be made. For instance, it emphasizes purposefulness and relevance rather than "utilitarian ends narrowly conceived for the masses." It proposed opportunities for learners to engage in problem-solving activities and learning experiences that "develop social insight and interest." This can be interpreted to mean that education in a democratic society must promote understanding and interest in societal issues among its recipients.

Earlier, Dewey had laid the curricular and psychological foundations for project-based instruction. In his paper titled *The Child and the Curriculum*, he noted that the "child is the starting point, the center and the end" (1902, p. 187) and that the most natural way for children to learn is by doing. However, they must be guided and provided with the appropriate learning experiences if they are to develop a habit of "critical examination and inquiry" (Dewey, 1910, p. 29). Although thinking is a natural process, Dewey insisted that we cultivate it in children through training. Dewey's work and ideas were extended by others in the progressive education movement, and by the latter part of the nineteenth century, there were several groups in the United States experimenting with different varieties of child-centered education.

In the early part of the twentieth century, among the personalities who popularized the idea of child-centered education was a professor at Teachers College by the name of William H. Kilpatrick. Kilpatrick had been studying projects as a method of teaching, and in 1918, he published an article in the *Teachers College Record* titled "The Project Method." In his article, Kilpatrick defines the project method, presenting a powerful case for the use of this method. In addition, he discusses different types of projects, their philosophical and social implications, and provided several examples of projects that students can do in the classroom. According to Spring (2001), Kilpatrick's article on the project method was so popular that it went through seven printings.

Psychological Foundations of the Project-based Approach

By the early to mid-1900s, the project method was given further theoretical capital by constructivists such as Jean Piaget and Lev Vygotsky (Howe & Berv, 2000). Piaget's research showed that children go through stages of cognitive development, although the ages at which they go through these stages varies (Piaget 1967, 1969, 1971). In addition, he demonstrated that knowledge acquisition and development takes place through a gradual process of interacting with the environment. Such an environment must be rich in materials, tools, and manipulatives to provide the child with the opportunity to construct and deconstruct his or her world. Vygotsky (1978) adds another dimension to Piaget's research by showing that learning is the driving force for children's cognitive development. According to him,

> Learning awakens a variety of internal development processes that are able to operate only when the child is interacting with people in his environment and in cooperation with his peers. Once these processes are internalized, they become part of the child's independent developmental achievement. (p. 90)

Vygotsky believes that a gap exists between what the child knows and can do on his or her own and what he or she knows and can do under adult guidance. He identifies this as the "zone of proximal development." By finding out where each child's zone of proximal development is, teachers can plan instruction to better meet their student's learning needs.

The works of Piaget and Vygotsky were ground breaking in the sense that they illuminated our understanding of the relationship among cognitive development, learning, and the environment. They also launched the constructivist movement, which promoted the idea that real knowledge is constructed through a process of social interaction, practice, and reflections. There are several constructivists who contributed to the development of project-based, student-centered learning and it is beyond the scope of this introduction to cover them all. However, it is appropriate to describe the contributions of Rosalind Driver and her colleagues.

Rosalind Driver was a lecturer in education and director of the Children's Learning in Science Project at the Center for Studies in Science and Mathematics Education, University of Leeds, UK. In a tribute to her, Osborne, Leach, and Scott (1997) wrote:

> Rosalind Driver was one of the pre-eminent figures in science education of her generation. She was a major presence on both the national and international stages, attracting interest and respect both from science education researchers and science teachers. Throughout her professional career she displayed an enduring passion for science education. She took very seriously the responsibility of seeking to improve our understanding of what is involved in teaching and learning science and, indeed, what might constitute an education in science. (p. 1)

The contributions of Rosalind Driver and her colleagues have great relevance to PBSI because it provides a solid bridge between how scientific knowledge is constructed and its relationship to learning science via project work. According to Driver, Asoko, Leach, Mortimer, and Scott (1994), scientific knowledge, regardless of discipline, consists of ideas and concepts that are constructed and "imposed on phenomena in attempts to interpret and explain them, often as results of considerable intellectual struggles" (p. 5). Such knowledge, when agreed upon by the scientific community, then becomes a "'taken-for-granted' way of seeing things within that community" (p. 5). Driver and her colleagues argue that the process of knowledge construction is both personal and social, and learning science

> requires more than challenging learners' prior ideas through discrepant events. Learning science involves young people entering into a different way of thinking about and explaining the natural world; becoming socialized to a greater or lesser extent into the practices of the scientific community with its particular purposes, ways of seeing, and ways of supporting its knowledge claims. Before this can happen, however, individuals must engage in a process of personal construction and meaning making. (p. 8)

This has implications for how we organize and implement science teaching and learning. Learning in a PBSI environment is well aligned to the above constructivist principles. In a PBSI environment, science learning is a social activity and students develop an understanding and appreciation of science, because they are immersed in a community where they learn the ideas, practices, and culture of science as it is practiced in the real world.

Another significant contribution to the constructivist worldview of science education that has had implications for PBSI and is worth noting in this introduction is the work of Mintzes, Wandersee, and Novak (1998). In their edited book titled *Teaching Science for Understanding: A Human Constructivist View*, the authors provide a framework for science educators for "making decisions about curriculum and instruction in the twenty-first century" (p. xviii). They propose a constructivist view of science teaching and learning that is neither radical nor social, but humanistic. According to Mintzes, Wandersee, and Novak,

> In contrast to the notion of radical and social constructivists, Human Constructivists take a moderate position on the nature of science. On the one hand, we find the views of classical "logical-positivists" intellectually indefensible; on the other, we think that many constructivists have created a relativistic mind-world that is ultimately self-defeating. We prefer instead a view of science that acknowledges an

external and knowable world, but depends critically on an intellectually demanding struggle to construct heuristically powerful explanations through extended periods of interaction with objects, events and other people. In its simplest form, human beings are meaning-makers; that the goal of education is to construct shared meanings and that this goal may be facilitated through the active intervention of well-prepared teachers. (p. xviii)

Mintzes, Wandersee, and Novak challenge the notion that there is only one approach to thinking about constructivist science teaching and learning. It not only provides a moderate view of constructivist science teaching and learning but also presents interesting and relevant ideas about learning science in PBSI and related inquiry-based science environments. These ideas, which are divided into theoretical and empirical foundations and theory-driven strategies of human constructivist approach to science education, were instrumental in shaping some of our current PBSI practices today.

It is fair to say that our knowledge of how students learn in the science classroom has improved tremendously during the past half a century, thanks to research on the relationships among the human brain, cognition, learning, and practical experience (Holt, 1983; Gardner, 1983, 1991; Novak, Cowin, & Johansen, 1983; Driver, Guesne, & Tiberghien, 1985; Carey, Evans, Honda, Jay, & Unger, 1989; Technology and Cognition Group at Vanderbilt, 1990; Novak, 1993; Papert, 1980, 1993; Tobin, 1993; Brown & Campione, 1994, 1996; Gunstone & Mitchell, 1998; National Research Council, 2000). From this research, we now know that all students have the capacity to learn science and that learning science is better when students are actively engaged in topics they can relate to their lived experiences, supported by caring adults in a safe and nurturing learning environment.

Project-based Science Instruction in the United States

The exact date when project-based science instruction started in the United States is difficult to determine. However, the use of the term "project" in science instruction can be traced back to 1908, when Rufus W. Stimson, a teacher at Smith Agricultural School in Northampton, Massachusetts, coined the term "home projects" (Stimson, 1914; Stevenson, 1928). The purpose of the "home projects" were to provide students with the opportunity to apply the teaching of the school in their farm work. In the 1920s and 1930s, project-based science instruction became very popular, particularly in the elementary and middle schools (Kilpatrick, 1924; Kliebard, 1986). Evaluation of project-based science instruction during this time showed that students who were taught using this method tended to do better on the Stanford Achievement Test, demonstrated better problem-solving skills and creativity, and read more books per year than students taught using traditional methods. In addition, teachers who used this method of teaching were found to be more enthusiastic about their work compared to their counterparts who used other methods of instruction (Hannes, 1921; Oberholtzer, 1934; Whipple, 1934; Aikin, 1942).

Despite its success, project-based science instruction did not spread as rapidly into high schools as it did in the elementary and middle schools. This was probably due to the fact that the high school curriculum was less flexible than the elementary and middle school curriculum. In addition, high school teachers were constantly under pressure to cover a certain amount of material and prepare their students for college admission. The opponents of project-based science instruction charged that this method of instruction focused too much on purposefulness and the learner's interest and ignored the importance of subject-matter knowledge (Charters, 1922; Bagley, 1921).

By the 1930s the debate between those who advocated a project-based, child-centered curriculum and those who advocated a subject-matter, objective-driven curriculum intensified. The latter group eventually dominated because, during this period, project-based, child-centered curriculums were considered to be too controversial (Kliebard, 1986). In addition, they did not fit well into the industrial model and function of schooling at that time (Spring, 1999). According to Polman (2000),

> Task structures, such as projects, however, can be at odds with the teachers' priority of maintaining order, as well as the students' priority of getting an optimal grade, because they increase ambiguity on how to perform to achieve a good grade and therefore increase perceived risk for the students. It is interesting to note that Dewey himself warned that "the mechanics of school organization and administration" (Dewey 1901, p. 337) often doomed reforms, but his warning did not prevent such factors from affecting his own efforts. (p. 27)

It is now more than half a century after the "great curriculum debate" and project-based science instruction has resurfaced again. Interest in project-based science instruction has been rekindled because of two possible factors: the standards movement in the 1980s (National Commission of Excellence in Education, 1983; American Association for the Advancement of Science, 1989) and increased National Science Foundation funding of programs and projects that promoted project-based science instruction in the 1990s. The standards movement, which began in the early 1980s, was sparked by *A Nation at Risk* (National Commission of Excellence in Education, 1983), a major national report on the condition of education in the United States. According to the report, American students were not performing up to expectations in the major subjects and therefore lacked basic academic skills necessary to compete with students in other industrialized countries. In order to remedy the situation, the report recommended the development and implementation of learning standards: what all students should know and be able to do.

Since the coming out of *A Nation at Risk*, several science education reports have been produced by various groups on the state of science education in the United States (National Commission on Mathematics and Science Teaching, 2000; Mullis, Martin, Beaton, Gonzalez, Kelly, & Smith, 1998; Schmidt, McKnight, & Raizen, 1997; National Education Goals Panel, 1997; Beaton, Martin, Mullis, Gonzalez, Smith, & Kelly, 1996; Raizen & Michelsohn, 1994; Mullis, Dossey, Campbell, Gentile, O'Sullivan,& Latham, 1994). Overall, these reports tended to paint a dim picture of the situation and recommended major reforms and more funding for science education. In addition, several scientific and professional teaching organizations (e.g., American Association for the Advancement of Science, National Science Foundation, and the National Science Teachers Association) responded by developing national science education standards (National Research Council, 1996), which have been adopted by most states and local education agencies in the United States.

Although the learning standards are not without controversy, they provide the foundation for the reform of science education and, in some cases, the impetus toward the implementation of inquiry-based or "hands-on" science-teaching methods in some U.S. classrooms. In general, the standards recognize the diversity in students' learning styles and emphasize the use of student-centered, inquiry-based methods of teaching science. The standards also call for the learning of science process skills and specific-content knowledge and for the use of performance-based assessments. The National Science Foundation (NSF), along with other philanthropic organizations, has, in the past two decades, funded a number of programs and projects aimed at reforming science teaching and learning. A major consequence of this targeted funding over the years has been the development and spread of a variety of project-based science curricula.

The work of Tinker and Papert (1989) deserves special attention here since the two scientists-researchers did not only demystify project-based science instruction but also popularized it in the science education community. Their research showed that the computer is a powerful tool in project-based science learning and can provide students with unique opportunities to construct models representing their own intellectual development (Tinker, 1991, 1992; Papert, 1980, 1993). According to them,

> Technology has something to offer every aspect of constructivist student activities. It can expand the range of possible projects; offer new opportunities for collaboration and communication; simplify acquiring and displaying of data; provide mechanisms to control experiments; increase the sophistication of theory-building, modeling, and data analysis; provide new outlets for creative expression; and grant access to vast databases of information. Without technology, practical issues of classroom management and limitations in the scope of potential student projects make student-centered activities difficult to offer and sustain. (Tinker & Papert, p. 6)

Tinker and Papert's research focused on the development and testing of microcomputer-based laboratories (MBL), which refers to the use of "microcomputers for student-directed data acquisition, display, and analysis" (p. 8) and LOGO, an interactive, object-oriented, programming language that allows learners to construct their own knowledge. They found that these computer-based tools enhance students' cognitive development in science. In addition, they expand the range of science problems to be investigated that otherwise would be limited due to funding constraints, scheduling, safety, time, and scale and make large-scale student collaboration on research projects possible.

Following the works of Tinker and Papert, Cohen (1997) edited a book titled *Internet Links for Science Education: Student-Scientist Partnerships*, which documents case histories of ten major project-based science curricula that have been implemented nationwide and how they impact the way students learn science. Most of the projects were funded by the NSF, which was interested in the feasibility of scaling up the implementation of project-based science curricula in K–12 settings as a way to promote better science teaching among teachers and deeper understanding of science content and process among students. The PBSI cases provided by Cohen offer great opportunities for teachers and their students to be involved with projects if they do not have the necessary resources or would like to try this instructional method first to learn more about it. Most of the projects presented usually recruit teachers and their students to participate in pilots and are eager to form partnership with local schools.

One organization that has been in the forefront of project-based science curriculum development is TERC, based in Cambridge, Massachusetts. With funding from the NSF, TERC has developed, tested, and disseminated several project-based science curricula for elementary, middle, and high schools. Two of their most successful and widely disseminated project-based science curricula are the Kids Network Project, now published by the National Geographic Society, and the Global Lab Project, published by Kendall-Hunt Publishers. The Kids Network Project is designed for elementary and middle school students, while the Global Lab Project is for secondary or high school students. Both curricula promote the teaching and learning of science using the project-based approach. The main idea behind these two projects is to build students' science process skills through collaborative scientific research that focuses on both local and global environmental problems. Participants include teachers and students from across the United States and abroad, as well as practicing scientists. Each participating school receives scientific tools for environmental research, access to an electronic community of learners, technology, and state-of-the-art written instructional materials.

The underlying thinking behind these project-based science curricula is that giving students access to a electronic community of learners and scientific tools will enable them to collect and share data on local and global environmental problems. In addition, they will build a community of students, teachers, and scientists with similar interests and that community of people with similar interests, guided by experienced teachers and scientists, will be able to create the social context essential for the authentic pursuit of scientific knowledge. Becoming part of this community will allow students to experience project-based science learning in a way that is meaningful, relevant, interdisciplinary, technology supported, and collaborative.

What Is Project-based Science Instruction?

Several scholars have attempted to define the term *project*. However, there are four definitions that have shaped our understanding of the term *project* and they are definitions by Kilpatrick (1918); Stockton; Hosic and Chase; and Stevenson. According to Kilpatrick, a project "is a wholehearted purposeful activity proceeding in a social environment" (p. 320). This definition implies two qualities: (1) that a project is a purposeful act and (2) that learning with projects is a social process where the learner interacts with people, places, and things.

Stockton's definition differs from Kilpatrick in that the term *project* is used to mean a "method." It is defined as the application of the principles of "self-education through activity" (p. 89). This means that a project is a learner-centered activity that requires the learner to take some responsibility for his or her own learning. In addition, it means that a project requires active participation rather than passive participation. As for Hosic and Chase, a project is a "complete, purposeful experience" (p. 11). The definition is similar to the definition offered by Kilpatrick. However, it differs in that it implies a learning experience or process with a beginning and an ending phase.

According to Stevenson, a project is "a problematic act carried to completion in its natural setting" (p. 43). This definition captures two of the elements already noted above as well as introducing a new element. The two elements already mentioned are: (1) "carried to completion," meaning that a project has a lifespan, that is, a beginning and an ending phase, and (2) that a project takes place in a "natural setting," which could be interpreted to mean a social environment or context. The new element that Stevenson's definition introduces is a "problematic act," which means that a project involves problem solving and critical thinking. Based on the above definitions, what then is a project? A project will be defined as a purposeful learner-centered activity with a definite life history and context, which involves problem solving and critical thinking conducted under the guidance of a teacher or mentor.

More recently, a small group of science education researchers have attempted to provide a working definition of the term *project-based science instruction*. These include Tinker (1991, 1992), Laffey, Tupper, Musser, and Wedman (1998), Krajcik, Czerniak, and Berger (1999), and Moje, Collazo, Carrillo, and Marx (2001). A review of each of their definitions of project-based science instruction is presented in table 1.1.

A close examination of table 1.1 indicates that Tinker's definition of project-based science instruction focuses on the scientific process. This definition indicates that science is learned best by doing science and that by doing science, one learns science process skills and content as well. The definition by Laffey et al. contains three elements: (1) contextual instruction, (2) student problem finding and framing, and (3) extended periods of time. This means that in a project-based science classroom, students learn by identifying, defining, and investigating problems over extended periods of time. In addition, the learning that takes place is grounded in problems that are relevant to students' own lives.

Table 1.1. Definition of Project-based Science Instruction.

Science Education Researchers	Definition of Project-based Science Instruction
Tinker (1992)	Projects are what scientists do. Students who are thoroughly engaged in a project, having selected the topic, decided on the approach, performed the experiment, drawn conclusion, and communicated the results, are doing science. They are seeing science not as a noun, an object consisting of facts and formulas, but as a verb, a process, a set of activities, a way of proceeding and thinking. (p. 33)
Laffey and colleagues	Project-based learning is a form of contextual instruction that places great emphasis on student problem-finding and framing, and which is often carried over extended periods of time. (p. 74)
Krajcik, Czerniak, and Berger	Project-based science has several fundamental features. First, driving questions or problems serve to organize and guide instruction. Second, students engaged in investigations to answer their questions. Third, communities of students, teachers, and members of society collaborate on question or problem. Fourth, students use technology to investigate, develop artifacts or products. Finally, the result is a series of artifacts or products that address the question or problem. (p. 9)
Moje and colleagues	Typically, the features of what is often called project-based pedagogy include (a) questions that encompass worthwhile and meaningful content anchored in authentic or real world problems; (b) investigations and artifacts creation that allow students to learn to apply concepts, represent knowledge, and receive ongoing feedback; (c) collaboration among students, teachers, and others in the community; and (d) use of literacy and technological tools. (p. 469)

Source: Colley, 2008, originally published in *The Science Teacher*, November 2008. (Copyright © National Science Teachers Association, all rights reserved.)

The definition by Laffey et al. is similar to Stevenson's definition and does not provide any new information. Krajcik et al.'s and Moje et al.'s definitions of project-based science instruction offer new information. They both include the following elements: (1) meaningful investigative questions or problems; (2) collaboration; (3) use of technology; and (4) production of artifacts. These two definitions do not indicate a time frame or context of instruction. Nevertheless, it can be assumed that meaningful investigations of scientific question, collaboration, and production of artifacts usually takes time and occurs within a particular context.

As one can see from the above definitions, the term *project-based science instruction* means different things to different people. However, the definitions also share some commonalities such as purposefulness, question driven, collaboration, use of tools and technology, artifacts, a context, and extended time frame. To sum up, project-based science instruction will be defined as a science teaching approach that is driven by well-defined student research questions and specific tangible outcomes. It is purposeful, collaborative, contextual, time-intensive science instruction that requires the use of appropriate tools and technology.

Categories of Projects

Kilpatrick (1918) identifies four categories of projects based on the purposes they serve. According to the author,

> Type 1 [project is], where the purpose is to embody some idea or plan in external form, as building a boat, writing a letter, presenting a play; type 2, where the purpose is to enjoy some (aesthetic) experience, as listening to a story, hearing a symphony, appreciating a picture; type 3, where the purpose is to straighten out some intellectual difficulty, to solve some problem, as to find out whether or not dew falls, to ascertain how New York outgrew Philadelphia; type 4, where the purpose is to obtain some item or degree of skill or knowledge, as learning to write grade 14 on the Thorndike Scale, learning the irregular verbs in French. (p. 332)

A review of the above quote suggests that there are three categories of projects instead of four. They can be summarized as follows:

Type 1 and Type 4. These projects are designed to teach students about a specific skill or set of skills and knowledge in a particular field of interest. I will, therefore, refer to them as *vocational skill projects* because their purpose is the development of lifelong skills.

Type 2. These projects are designed to make students aware of and appreciate certain fields, concepts, knowledge, and attitudes. I will, therefore, refer to them as *awareness and appreciation* projects.

Type 3. These projects are designed to provide students with process skills, such as the ability to think critically, to problem solve, to collect and analyze data, to interpret data, and to draw conclusions. I will, therefore, refer to them as *process skills* projects.

Other than purpose, projects can also be categorized by subject matter: biology, physics, chemistry, history, geography, mathematics, languages, arts, and so on; by duration: short term, medium term, or long term; by degree of utility: applied or theoretical; by scale: small scale, large scale, pilot, extensive, exploratory, or in-depth; and by method of implementation: experimentation, observation, interview, survey, or mixed method. These categorizations of projects will not suit every project; as a result, new categorizations will have to be developed when the need arises.

It is important for teachers who intend to implement project-based science instruction to understand the different categories of projects so they can guide their students in the selection of their projects. Without a clear understanding of what type of project to conduct, it will be difficult for students to plan, implement, and evaluate the outcomes of their projects.

Science Learning Standards and Project-based Science Instruction

In 1983, the National Commission on Excellence in Education issued a report titled *A Nation at Risk: The Imperative for Educational Reform*. In that report, the Commission presented findings in four areas of education, namely: content, expectations, time, and teaching. Speaking about teaching, the Commission

> found that not enough of the academically able students are being attracted to teaching; that teacher preparation programs need substantial improvement; that the professional working life of teachers is on the whole unacceptable; and that a serious shortage of teachers exists in key fields. (p. 20)

In response to *A Nation at Risk*, the American Association for the Advancement of Science (1989) released a publication titled *Project 2061: Science for All Americans*. According to the Association,

> Science education—meaning education in science, mathematics, and technology—should help students to develop the understandings and habits of mind they need to become compassionate human beings able to think for themselves and to face life head on. It should equip them also to participate thoughtfully with fellow citizens in building and protecting a society that is open, decent and vital. (p. xiii)

Similarly, the National Research Council (1996) released the *National Science Education Standards* (NSES), which presented a vision for science education in the United States in the twenty-first century. The standards focused on all the things students and teachers should know, understand, and be able to do, and covered science content, science teaching, assessment, science education programs, professional development, and the science education system. The release of *Project 2061* and the NSES provided a new framework for science teaching and learning. In addition, it created new opportunities and challenges for preparing science teachers.

Since the introduction of *A Nation at Risk*, many states have developed and implemented their own learning standards, which are aligned to national and professional standards. Schools have planned and implemented professional development to help teachers prepare students to meet the standards.

According to the NSES, the most appropriate way to teach students science is through the inquiry-based method. This means having students engage in investigating science questions or solve science-related problems that are tied to content and curriculum standards. The inquiry-based method requires students and teachers to move away from the traditional lecture and lab method and engage in messy activities that may not always be predictable. However, although there is more awareness about learning standards and inquiry-based science learning, it is still not the dominant teaching method used in science classrooms in the United States. There are several reasons for this.

One reason is related to contextual factors in some schools. It is not uncommon to find poorly resourced schools in urban and rural America where facilities for teaching science are grossly inadequate or antiquated. Under such conditions, it will be very difficult or impossible to implement inquiry-based science instruction. Another reason is related to teacher preparation in the inquiry-based science and related instructional methods. From my experience as a science teacher-educator for over two and half decades, I have observed that not all preservice science teacher preparation programs in the United States offer courses or training specifically on how to conduct inquiry-based science in the classroom. Usually, inquiry-based science instruction is covered as part of a science education method course and may or may not be adequate or rigorous. Some science teachers actually learn about inquiry-based science instruction while in-service when they formally become full-time employed teachers. As a result, they and their students learn by trial and error.

Another reason is related to conception and misconception about inquiry-based science instruction. It is one of those methods of science instruction with a popular name, but not well understood. Anderson (2002) summarized the challenge of preparing science teachers in inquiry-based science instruction as follows:

> The task of preparing science teachers for inquiry teaching is much bigger than the technical matters. Even though teachers need to learn how to teach students constructively, acquire new assessment competencies, learn new teaching roles, learn how to put students in new roles, and foster new forms of students' work, the task of preparing teachers for inquiry teaching includes much more. (p. 8)

In 2013, a new set of science learning standards called the *Next Generation Science Standards* (NGSS) were released. The NGSS were developed in partnership with the National Research Council (NRC); the National Science Teachers Association (NSTA); the American Association for the Advancement of Science (AAAS); Achieve, an independent, bipartisan education reform group; and 26 lead states (Next Generation Science Standards Lead States, 2013; also see www.nextgenscience.org/lead-state-partners). The first step was the development of the *Framework for K–12 Science Education* (National Research Council, 2012), which documents the basic ideas and practices in science and engineering that all students should know and be able to do by the time they graduate from high school. The second step was the development of standards based on the *Framework*. During the second step, the 26 lead states provided leadership to the standards development teams and made a commitment to implement them. Currently, there are different professional development activities and initiatives being implemented in the lead states to promote the use of the NGSS standards.

According to the Next Generation Science Standards Lead States, "Every NGSS standard has three prongs: content, scientific and engineering practices and cross-cutting concepts. The integration of rigorous content and application reflects how science is practiced in the real world" (p. 1). Both the NSES and the NGSS call for a fundamental shift in how science is conducted in the classroom. In order for K–12 science teaching and learning to change and reflect how science is "practiced in the real world," our thinking about the nature of science, the instructional methods, assessment methods, curriculum materials, tools, and technology we use, and how we organize science learning environments, needs to change to allow teachers and students to be actively engaged in science learning. PBSI is well aligned to the NSES and NGSS because it reflects how science is conducted in the real world. When students are engaged in project work, not only do they learn science content, but they also acquire science process skills (i.e., scientific and engineering practices), learn how to collaborate, engage in interdisciplinary learning (i.e., cross-cutting concepts), and build a scientific community of practice.

According to Tinker, (1991) PBSI comes with many practical advantages over traditional instructional methods such as:

1. adaptability to learners with different learning styles,
2. interdisciplinarity because they require the application of different disciplinary knowledges, skills, and dispositions,
3. integrative, meaning that PBSI allows students to see how things are connected and interconnected,
4. pre-professional, meaning that project work mirrors the world of work,
5. motivational, because students come up with their own research questions to study and take ownership of their own learning,
6. effective, because knowledge is constructed by the students themselves in partnership with their teachers, and
7. efficient, because it allows for and promotes the use of tools and technology in the learning process.

All of these characteristics of PBSI are consistent with the *Framework for K–12 Science Education* recommendations, are well aligned to the NSES and NGSS, and provide a powerful option for science teachers and their students to learn science. As Metz (2015) acknowledges,

> Project-based science can be an important instructional model for meeting the three-dimensional learning goals of the Next Generation Science Standards. Complex real-world projects provide opportunities for students to deeply engage in multiple science and engineering practices—like developing and using models, constructing explanations, and engaging in arguments for evidence, while learning specific disciplinary core ideas and crosscutting concepts. (p. 6)

Chapter Summary

In this introduction to PBSI, we have learned that the underlying principles of project-based science instruction are deeply rooted in the child-centered educational movement of seventeenth-century Europe, which was transferred to the United States and promoted by the progressive education movement. Although project-based science instruction did not enjoy long-term support after its introduction, it is enjoying a revival today due to increased funding from the National Science Foundation and other nongovernmental educational organizations and increased knowledge about how projects work in the science classroom coming out of the science education research field.

In addition, we have also learned that the main defining quality of PBSI as a science instructional method is that it is driven by well-defined student research questions and specific tangible outcomes. It is purposeful, collaborative, contextual, time-intensive science instruction that requires the use of appropriate tools and technology. PBSI is consistent with how science is practiced in the real world and is a powerful instructional strategy for achieving state and national science learning standards.

Food for Thought

The following questions are to help the reader reflect on the chapter.

1. Based on this chapter, what do you think are the contextual factors that led to the development of PBSI in the United States?
2. What are the critical elements of PBSI?

· 2 ·

WHAT DOES RESEARCH SAY ABOUT PROJECT-BASED SCIENCE INSTRUCTION?

Chapter Overview

Although the current body of research on PBSI has not peaked yet, it is nevertheless growing steadily. In the past two decades, some interesting studies have been conducted on PBSI covering a wide range of topics, themes, questions, and problems. A review of this body of research on PBSI will be conducted with particular emphasis on peer-reviewed or refereed studies.

Conducting the Review

In order to answer the following question, What does the research says about PBSI?, the method of research synthesis (Sadler, Burgin, McKinney, & Ponjuan, 2010; Labin, Duffy, Mayers, Wandersman, & Lesesne, 2012) was used. The sampling strategy of choice was purposeful sampling because this allows only studies directly relating to PBSI that were published in peer-reviewed journals to be sampled. In addition, peer-reviewed articles gathered over time by the author on the PBSI (including the author's own) were included in the sample. An extensive search was conducted online using Google's search engine, online educational databases, and related search tools. Searches were conducted on archived and current issues of the following journals: *Science Teacher*, *Science*

Education, Journal of Research in Science Teaching, Electronic Journal of Science Education, International Journal of Science Education, International Journal of Science and Mathematics Education, Research in Science Education, Science Educator, Science Education International, American Biology Teacher, Journal of Chemical Education, and *Journal of College Science Teaching*. To identify and collect documents, the following search terms were used: *project-based, project-based science, project-based science instruction, project-based science learning,* and *project-based science teaching*. Out of more than 300 documents identified, 71 met the criteria for inclusion in this review. The criteria for inclusion in the review were as follows: (1) peer-reviewed or refereed, (2) published between 1990–2015, and (3) included content that directly relates to PBSI.

Table 2.1 shows a list of the research articles reviewed for this chapter. The list is organized in chronological order beginning with the most recent articles published in the current year and ending with the earlier or oldest ones published in the early 1990s. For analysis purposes, the table is partitioned into the following variables: author(s), study titles, disciplines, N or sample size, duration of study, methods of study used, and classifications. By including the author(s) names, it is possible to identify who PBSI researchers, scholars, and practitioners are and from where the research is originating. For instance, by reviewing the articles, it is clear that Dr. Joseph Krajcik and his colleagues at the University of Michigan have authored most of the empirical research articles on PBSI in the United States and contributed the most articles per capita. It is very important to acknowledge their valuable contributions to the field of PBSI research and for being pioneers in this important area of science education research. The study titles are included in the table to demonstrate the diversity of PBSI research and provide the reader who is unfamiliar with or may not be able to read all the articles a quick sense of what PBSI research is going on and what topics are being investigated or studied. The column "discipline" refers to the specific content area the articles address. The review shows that about 12 disciplines and/or cross-disciplines are represented. These include astronomy, biology, chemistry, earth science, earth and space science, environmental science, engineering, environmental chemistry, physics, STEM (science, technology, engineering, and mathematics), technology, and PBSI. "PBSI research articles" refers to articles that focus on curriculum development and evaluation, instruction, and assessment of students, and/or teacher professional development in PBSI environments. Of the 71 research articles identified, 29 were on PBSI, 11 were in chemistry, 9 were in biology, 6 in physics, 4 in engineering, 2 in STEM, and 2 in technology.

The rest were cross-disciplinary/interdisciplinary consisting of environmental science, earth and space science, environmental chemistry, biology, and technology. Only one article was on astronomy.

In conducting the review, it was important to identify the sample of students and teachers (N) who participated in the reported studies in order to get a rough sense of potential impact and those exposed to PBSI at the K–16 levels. An examination of table 2.1 shows that the "N" column has two configurations of numbers in the form number-slash-number (0/1). The first number refers to sample of students and second number refers to sample of teachers. For example, looking at Entry 12, Cook, Buck, and Rogers (2012), the N reads 30/1. This means that the study reported 30 students and one teacher. When only one number is shown, as in Entry 1, Han, Capraro, and Capraro (2015), where N equals 1,890, this refers only to the sample of students who participated in that study. As can be seen, N varies greatly from 8 to 7,997 for students and 1 to 102 for teachers. Based on the 71 studies, approximately, 16, 936 students and 704 teachers were reported to have taken part in PBSI research-related activities. An important caveat to note, though, is that 15 of the articles in the review did not report N or this piece of information was left out. This means that the total numbers of students and teachers reported to participated in PBSI research-related activities could be higher.

The duration of study refers to how long a particular study was conducted and this varies from hours to years. Some studies report duration in lessons, days, weeks, months, semesters, and years. The most common durations of study that were identified in the review were those that were implemented on a weekly basis, ranging from 3 to 18 weeks, and semester basis, ranging from one to four semesters. Seventeen articles in the review did not report duration or this piece of information was left out.

The last column on table 2.1 is the class or classification and it refers to categories in which articles are grouped. To determine the research traditions or paradigms PBSI researchers applied the most and least, the following classification system was created to organize the review:

1. Empirical research article (ERA), an article based on an explicit research question/hypothesis and/or problem, with a clear methodology and emphasis on data/results/evidence to support findings.
2. Evaluation research article (EVRA), an article on PBSI program or project implementation and how goals and/or objectives were achieved or not achieved.

3. Theoretical research article (TRA), a descriptive, analytical, or critical essay or review on a particular area, question, or problem that is based on opinion, experience, and/or theory.
4. Action research article (ARA), an article on classroom practices or instructional innovation conducted by teachers-practitioners to improve practice.

After a thorough examination of the 71 articles, 34 were classified as ERA, 14 as TRA, 15 as EVRA, and 8 as ARA.

What Does the Research Says about PBSI?

An analysis of the research articles on PBSI in this review reveals the following themes: developing and evaluating PBSI curriculum materials, engagement of teachers on PBSI professional development, engagement of students in project-based science learning, impact on student learning, and applications of PBSI within and across science disciplines. Each theme will be examined and summarized in terms of what the research says.

PBSI Curriculum Materials

Eight articles focused on developing and evaluating PBSI curriculum materials (Duncan & Tseng, 2011; Kanter & Konstantopoulos; Kanter, 2010; Krajcik, McNeill, & Reiser, 2008; Colley, 2008; Schneider, Krajcik, & Blumenfeld, 2005; Rivet & Krajcik, 2004; Barron, Schwartz, Vye, Moore, Petrosino, Zech, & Bransford, 1998). An examination of these PBSI curricula shows that they all share some elements in common, such as having an explicit design framework, alignment to science learning standards, relevance to students' lives, use of technology, and professional development-training for the teachers. It is interesting to note that some of the PBSI curricula were field-tested in large urban school districts with large sample sizes, sometimes over extended periods. Students who participated in these PBSI curricula were reported to have demonstrated positive outcomes and performed better on a national measure of science achievement (Schneider Krajcik, Marx, & Soloway, 2002). Three of the studies indicated that PBSI has a positive effect on content knowledge and understanding of science, especially for traditionally low achievers and underrepresented students (Duncan & Tseng; Kanter & Konstantopoulos; Marx Blumenfeld, Krajcik, Fishman, Soloway, Geier & Tal, 2004). Although

WHAT DOES RESEARCH SAY ABOUT PROJECT-BASED SCIENCE? 29

Table 2.1. Summary of Research on PBSI.

#	Author(s)	Study Title	Discipline	N[1]	Duration[2]	Methods[3]	Class[4]
1	Colaianne (2015)	"Global Warming: Project-based Science Inspired by the Intergovernmental Panel on Climate Change"	Earth & Space Science	UNK	4 weeks	The Intergovernmental Panel on Climate Change (IPCC) used as a framework for organizing PBSI, peer review, and presentations	ARA
2	Han, Capraro, & Capraro (2015)	"How Science, Technology, Engineering, & Mathematics (STEM) Project-based Learning (PBL) Affects High, Middle, and Low Achievers Differently: The Impact of Student Factors on Achievement"	STEM	1890	3 years	Quasi-experimental with treatment and control groups, Hierarchical Linear Model (HLM)	ERA
3	D'Amico, Gomez, & McGee (2015)	A Case Study of Student and Teacher Use of Projects in a Distributed Multimedia Learning Environment	Technology PBSI	275/6	1 year	Surveys, interviews, observations, e-mail, reports, and software logs	EVRA
4	Lee, Lee, & Lee (2015)	"Project Approach in South Korea: Kimchi/Kimjang"	PBSI	UNK	UNK	Before and after drawings; concept map	ERA
5	Krajcik (2015)	"Project-based Science: Engaging Students in Three-Dimensional Learning"	PBSI	N/A	N/A	Essay/review	TRA
6	Hike & Beck-Winchatz (2015)	"Near-Space Science: A Ballooning Project to Engage Students With Space Beyond the Big Screen"	Chemistry	80	1 semester	Student-centered experiments, science-writing heuristic, peer review, presentations	ARA
7	Cook & Weaver (2015)	"Teachers' Implementation of Project-based Learning: Lessons from the Research Goes to School Program"	PBSI	0/7	2 weeks	Case study: video recordings of observations and semi-structured interviews	ERA

#	Author(s)	Study Title	Discipline	N[1]	Duration[2]	Methods[3]	Class[4]
8	Rittenburg, Miller, Rust, Esler, Kreider, Boylan, & Squires (2015)	"The Community Connection: Engaging Students and Community Partners in Project-based Science"	Biology & Environmental Science	UNK	1 year	The Confluence Project (TCP) as a model for PBSI, peer review and presentations, students' pre- and postsurvey	ARA
9	de los Santos, Montes, Sánchez-Coronilla, & Navas (2014)	"Sol-gel Application for Consolidating Stone: An Example of Project-based Learning in a Physical Chemistry Lab"	Chemistry	21	24 hours	Essay/review	EVRA
10	Liu (2014)	*Implementing Project-based Learning in Physics and Statics Courses*	Physics	9	1–2 semesters	Surveys	EVRA
11	Bilgin, Karakuyu, & Ay (2014)	"The Effects of Project-based Learning on Undergraduate Students' Achievement and Self-efficacy Beliefs Towards Science Teaching"	PBSI	66	27 hours (3 hrs/week)	Pretest/posttest quasi-experimental method with a control group	ERA
12	Brickman, Gormally, Francom, Jardeleza, Schutte, Jordan, & Kanizay (2012)	"Media-savvy Scientific Literacy: Developing Critical Evaluation Skills by Investigating Scientific Claims"	Biology	UNK	UNK	Essay/review	TRA
13	Kiefer, Bucholtz, Goode, Hugdahl, & Trogden (2012)	"Undesired Synthetic Outcomes During a Project-based Organic Chemistry Laboratory Experience"	Chemistry	UNK	UNK	Essay/review	TRA
14	Cook, Buck, & Park Rogers (2012)	"Preparing Biology Teachers to Teach Evolution in a Project-based Approach"	Biology	30/1	3 weeks	Case study: audio and video tape, artifacts of students' work, field notes, teacher documents, students assessments, journals, and student interviews	ERA

WHAT DOES RESEARCH SAY ABOUT PROJECT-BASED SCIENCE? 31

15	Sadeh & Zion (2012)	"Which Type of Inquiry Project Do High School Biology Students Prefer: Open or Guided?"	Biology	295/18	2 years	Quasi-experiment, comparison groups, no control, student attitudinal questionnaire	ERA
16	Shen, Jensen, Wentz, & Fischer (2012)	"Teaching Sustainable Design Using BIM and Project-based Energy Simulations"	Engineering	12	16 weeks	Quizzes, assignments, graded homework, presentations; student products, surveys	EVRA
17	Wilson, Parkin, & Thomas (2011)	Frontiers of Crystallography: A Project-based Research-led Learning Exercise	Chemistry	20–30	4.5–5 hours	Essay/review	TRA
18	Park Rogers, Cross, Gresalfi, Trauth-Nare, & Buck (2011)	First Year Implementation of a Project-based Learning Approach: The Need for Addressing Teachers' Orientations in the Era of Reform	PBSI	0/3	1 year	Collective case study, grounded theory, interviews, videotapes of classroom activity, and teaching philosophy questionnaire	ERA
19	Sheikh, Fulbright, & Hademenos (2011)	"Captain R. Rubber Ducky: A STEM-Driven Project in Aquatic Robotics"	STEM	UNK	UNK	Descriptive report	EVRA
20	Palmer & Hall (2011)	"An Evaluation of a Project-based Learning Initiative in Engineering Education"	Engineering	72	1 semester	Questionnaire, Fisher's test, and ANOVA	EVRA
21	Toolin & Watson (2010)	"Students for Sustainable Energy: Inspiring Students to Tackle Energy Projects in Their School and Community"	Physics	43	1 semester	Performance assessment, self-assessment, peer assessment, open-ended surveys, and presentations	ARA

#	Author(s)	Study Title	Discipline	N[1]	Duration[2]	Methods[3]	Class[4]
22	Duncan & Tseng (2010)	"Designing Project-based Instruction to Foster Generative and Mechanistic Understandings in Genetics"	Biology	111	5 weeks	Pre- and postwritten assessments, pre- and post clinical interviews, artifacts of students' work, and videos of classroom instruction	ERA
23	Kosinski-Collins & Gordon-Messer (2010)	"Using Scientific Purposes to Improve Student Writing & Understanding in Undergraduate Biology Project-based Laboratories"	Biology	N/A	N/A	Essay/review	TRA
24	Alozie, Eklund, Rogat, & Krajcik (2010)	"Genetics in the 21st Century: The Benefits & Challenges of Incorporating a Project-based Genetics Unit in Biology Classrooms"	Biology	UNK	UNK	Essay/review	TRA
25	Kanter & Konstantopoulos (2010)	"The Impact of a Project-based Science Curriculum on Minority Student Achievement, Attitudes, and Careers: The Effects of Teacher Content and Pedagogical Content Knowledge and Inquiry-based Practices"	PBSI	301/9	10–12 weeks	Students pre- and posttests (achievement items) and surveys; teacher pre- and posttests (content knowledge and pedagogical content knowledge)	ERA
26	Kanter (2010)	"Doing the Project and Learning the Content: Designing Project-based Science Curricula for Meaningful Understanding"	PBSI	652/12	1 year	Pre- and posttests	ERA

27	Cook (2009)	"A Suggested Project-based Evolution Unit for High School: Teaching Content Through Application"	Biology	UNK	5 weeks	Essay/review	TRA
28	Bhattacharyya & Bhattacharya (2009)	"Technology-integrated Project-based Approach in Science Education: A Qualitative Study of In-service Teachers' Learning Experiences"	Technology	70	4 semesters	Observations, journal reflections, document analysis, and in-depth, open-ended interviews	ERA
29	Regassa & Morrison-Shetlar (2009)	"Student Learning in a Project-based Molecular Biology Course"	Biology	50	4 semesters	Pre- and post knowledge assessment, pre- and post focus group interviews, writing samples, critical thinking/problem-solving assessment, and self-confidence surveys	EVRA
30	Bencze & Bowen (2009)	"Student-teachers' Dialectically Developed Motivation for Promoting Student-led Science Projects"	PBSI	0/9	18 weeks	Video recordings, evaluations of propositions about science, semi-structured interviews, and samples of students' projects	ERA
31	Colley (2008)	"Project-based Science: A Primer—An Introduction and Learning Cycle for Implementing Project-based Science"	PBSI	N/A	N/A	Essay/review	TRA
32	Dickinson & Jackson (2008)	"Planning for Success: How to Design and Implement Project-based Science Activities"	PBSI	N/A	N/A	Essay/review	TRA

#	Author(s)	Study Title	Discipline	N[1]	Duration[2]	Methods[3]	Class[4]
33	Fallik, Eylon, & Rosenfeld (2008)	"Motivating Teachers to Enact Free-choice Project-based Learning in Science and Technology (PBLSAT): Effects of a Professional Development Model"	PBSI	0/65	28-56 hours	Before and after questionnaires and interviews	ERA
34	Weizman, Schwartz, & Fortus (2008)	"The Driving Question Board: A Visual Organizer for Project-based Science"	PBSI	N/A	N/A	Essay/review	TRA
35	Short, Lundsgaard, & Krajcik (2008)	"How Do Geckos Stick? Using Phenomena to Frame Project-based Science in Chemistry Classes"	Chemistry	N/A	N/A	Essay/review	ARA
36	Chin & Chia (2008)	"Problem-based Learning Tools: Problem-based Learning Pedagogy and Strategies Are Used to Implement Project-based Science"	PBSI	39	16 weeks	Problem log, group problem statement, need-to-know worksheet, project planner form, project tasks allocation form, survey questionnaire and interview schedule, learning log, self-evaluation form, and assessment rubric	ARA
37	Vacchina & Aguirre (2008)	"The Herpetology Project: Students Construct Traps to Collect and Analyze Turtle Data"	Environmental Chemistry, Environmental Science, & Biology	10/2	4 years	The Herpetology Project, Earth Watch Institute as model and field expedition to Brazil	ARA

38	Krajcik, McNeill, & Reiser (2008)	"Learning-goals-driven Design Model: Developing Curriculum Materials That Align with National Standards and Incorporate Project-based Pedagogy"	PBSI	870/11	2 years	Student pre- and posttests, student artifacts, field notes, classroom videos, teacher feedback, content expert feedback, and *Project 2061* review	ERA
39	Sola & Ojo (2007)	"Effects of Project, Inquiry and Lecture Demonstration Teaching Methods on Senior Secondary Students' Achievement in Separation of Mixtures Practical Test"	Chemistry	233	6 weeks	Quasi-experimental design with a control group, pre- and posttest	ERA
40	Tsaparlis & Gorezi (2007)	"Addition of a Project-based Component to a Conventional Expository Physical Chemistry Laboratory"	Chemistry	27	UNK	Written questionnaire	EVRA
41	Pearce (2007)	"Teaching Physics Using Appropriate Technology Projects"	Physics	~150	UNK	Course evaluations, pre- and postexam grades, and observation	EVRA
42	Wilhelm, Thacker, & Wilhelm (2007)	"Creating Constructivist Physics for Introductory University Classes"	Physics	38	UNK	Mixed method research design, students' final projects and presentations, pre- and post-Force Concept Inventories (FCI), and end-of-course interviews	ERA
43	Adami (2006)	"A New Project-based Lab for Undergraduate Environmental and Analytical Chemistry"	Chemistry	55	4 years	Course evaluation and questionnaire	EVRA

#	Author(s)	Study Title	Discipline	N[1]	Duration[2]	Methods[3]	Class[4]
44	Lehman, George, Buchanan, & Rush (2006)	"Preparing Teachers to Use Problem-centered, Inquiry-based Science: Lessons from a Four-year Professional Development Project"	PBSI	700/23	1 year	Pre- and post test, surveys, portfolios, observations, and informal interviews	ERA
45	Colley (2006)	"Understanding Ecology Content Knowledge and Acquiring Science Process Skills Through Project-based Science Instruction"	Earth Science	18	2 days	Pre- and post assessment	ERA
46	Tali, Krajcik, & Blumenfeld (2006)	"Urban Schools' Teachers Enacting Project-based Science"	PBSI	0/2	15 lessons	Qualitative interpretative study, videotaped lessons	ERA
47	Wu & Krajcik (2006)	"Exploring Middle School Students' Use of Inscriptions in Project-based Science Classrooms"	PBSI	27/2	18 weeks	Naturalistic approach, video recordings, field notes, students' science reports, computer-based models, digital pictures, web pages, and notebooks	ERA
48	Wilhelm & Confrey (2005)	"Designing Project-enhanced Environments: Students Investigate Waves and Sound"	Physics	9	UNK	Pre- and posttest, group projects	ARA
49	Barak & Dori (2005)	"Enhancing Undergraduate Students' Chemistry Understanding Through Project-based Learning in an IT Environment"	Chemistry & Technology	215	1 semester	Quasi-experiment, non-randomized, treatment and control groups, pre- and posttest, and final exams, content analysis, interviews, and observations	ERA

WHAT DOES RESEARCH SAY ABOUT PROJECT-BASED SCIENCE? 37

50	Schneider, Krajcik, & Blumenfeld (2005)	"Enacting Reform-based Science Materials: The Range of Teacher Enactments in Reform Classrooms"	PBSI	0/4	2 semesters	Qualitative case study, videotapes of classroom activities	ERA
51	Doppelt (2005)	"Assessment of Project-based Learning in a MECHATRONICS Context"	Engineering	54/18	UNK	Qualitative	ERA
52	Colley (2005)	"Project-based Science Instruction: Teaching Science for Understanding"	PBSI	N/A	N/A	Essay/review	TRA
53	Tal & Argaman (2005)	"Characteristics and Difficulties of Teachers Who Mentor Environmental Inquiry Projects"	Environmental Science	15	2 years	Interpretive research method, grounded theory, journals, questionnaires, interviews, and observations	ERA
54	Walton & Archer (2004)	"The Web and Information Literacy: Scaffolding the Use of Web Sources in a Project-based Curriculum"	Technology	UNK	3 years	Case study	ERA
55	Rivet & Krajcik (2004)	"Achieving Standards in Urban Systemic Reform: An Example of a Sixth Grade Project-based Science Curriculum"	PBSI	2,500/24	4 years	Pre- and posttest	ERA
56	Marx, Blumenfeld, Krajcik, Fishman, Soloway, Geier, & Tal (2004)	"Inquiry-based Science in the Middle Grades: Assessment of Learning in Urban Systemic Reform"	PBSI	7,997/102	3 years	Pre- and posttest	ERA

#	Author(s)	Study Title	Discipline	N[1]	Duration[2]	Methods[3]	Class[4]
57	Freeman, Marx, & Cimellaro (2004)	"Emerging Considerations for Professional Development Institutes for Science Teachers"	PBSI	0/58	11 days	Focus group interviews, individual reflections, pre- and postsession surveys, and observations	ERA
58	Draper (2004)	"Integrating Project-based Service-Learning into an Advanced Environmental Chemistry Course"	Chemistry	8	13 weeks	Corse evaluations	EVRA
59	Mills & Treagust (2003)	"Engineering Education—Is Problem-based or Project-based Learning the Answer?"	Engineering	UNK	UNK	Essay/review	EVRA
60	Schneider, Krajcik, Marx, & Soloway (2002)	"Performance of Students in Project-based Science Classrooms on a National Measure of Science Achievement"	PBSI	142	UNK	Quasi-experiment using two comparison groups, NAEP Assessment	ERA
61	Wright & Boggs (2002)	"Learning Cell Biology as a Team: A Project-based Approach to Upper-Division Cell Biology"	Biology	UNK	UNK	Before and after course grades	EVRA
62	Moje, Collazo, Carrillo, & Marx (2001)	"Maestro, What Is 'Quality'?": Language, Literacy, and Discourse in Project-based Science"	PBSI	0/1	UNK	Participant observation field notes, interviews, student writings, curriculum work sheets, audiotapes and videotapes of classroom activities, and electronic discussion	ERA

63	Evans, Abrams, Rock, & Spencer (2001)	"Student/Scientist Partnerships: A Teacher's Guide to Evaluating the Critical Components"	PBSI	UNK	UNK	Essay/review	TRA
64	Milner-Bolotin & Svinicki (2001)	"Teaching Physics of Everyday Life: Project-based Instruction and Collaborative Work in Undergraduate Physics Course for Nonscience Majors"	Physics	42	1 semester	Videotaped and analyzed multiple interviews, minute papers, student reflections, e-mail communications, end-of-activity open-ended questions, and end-of-project essays	ERA
65	Barab, Hay, Barnett, & Keating (2000)	"Virtual Solar System Project: Building Understanding Through Model Building"	Astronomy	63	2 years	Naturalistic inquiry using quantitative and qualitative data	ERA
66	Thomas (2000)	"A Review of Research on Project-based Learning"	PBSI	N/A	N/A	Literature review	EVRA
67	Crawford, Krajcik, & Marx (1999)	"Elements of a Community of Learners in a Middle School Science Classroom"	PBSI	26/1	12 weeks	Videotapes, interviews, teacher's journal, and electronic correspondence	ERA
68	Barron, Schwartz, Vye, Moore, Petrosino, Zech, & Bransford (1998)	"Doing with Understanding: Lessons from Research on Problem- and Project-based Learning"	PBSI	UNK	UNK	Synthesis and case studies	TRA
69	Krajcik, Blumenfeld, Marx, Bass, Fredricks, & Soloway (1998)	"Inquiry in Project-based Science Classrooms: Initial Attempts by Middle School Students"	PBSI	8	7 months	Case studies	ERA

#	Author(s)	Study Title	Discipline	N[1]	Duration[2]	Methods[3]	Class[4]
70	Juhl, Yearsley, & Silva (1997)	"Interdisciplinary Project-based Learning Through an Environmental Water Quality Study"	Chemistry	UNK	UNK	Surveys and observations	EVRA
71	Berenfeld (1994)	"Technology and the New Model of science Education: The Global Lab Experience"	Environmental Science & Technology	UNK	5 years	Descriptive report	TRA

Key to table

1 = "Sample" size or the number of students and teachers who participated in the study, whereas the first number refers to students and second number refers to teachers (e.g., 10/1 means ten students and one teacher). When only one number is shown, this refers only to the sample of student participants.

2 = "Duration" of study refers to how long the study was conducted; this varies from hours to years.

3 = "Methods" refers to methods of study and/or data collection and data analysis procedures used to conduct the study (e.g., experimental, quasi-experimental, case study, phenomenological, ethnography, grounded theory, naturalistic inquiry, survey, correlational study, mixed methods, questionnaire, pre- and posttests, interviews, observations, etc.).

4 = "Class" stands for classification and refers to categories the articles are grouped into, such as empirical research article (ERA), evaluation research article (EVRA), theoretical research article: opinion, essay, or review (TRA), and action research article (ARA).

"UNK" refers to unknown.

"N/A" refers to not applicable

these PBSI curricula were reported to be effective in engaging students and teachers in authentic science learning, the importance of training and support for teachers, as well as the policy and administrative environments in schools, were emphasized as factors essential to ensure their successful implementation. An important point to note is that most of the PBSI curricula reported in this review, with the exception of Berenfeld, were for the middle grades and little is known about implementation of PBSI curricula on a longitudinal basis at the high school level.

Engagement of Teachers in PBSI Professional Development

One of the major challenges in implementing PBSI in the science classroom is the lack of teacher knowledge and experience in this approach to science teaching (Colley, 2005). Some researchers have tried to address this challenge by providing training and professional development on PBSI and studying teachers' level of engagement (Cook & Weaver; Park Rogers et al.; Kanter & Konstantopoulos; Bencze & Bowen, 2007; Lehman et al.; Tal, Krajcik, & Blumenfeld; Colley, 2005; Freeman, Marx, & Cimellaro). With the exception of the studies by Lehman et al. and Freeman, Marx, and Cimellaro, which used mixed methods and sample sizes of 23 and 58 respectively, all the other studies are qualitative in methodology and used a small sample of teachers that varies from 3 to 15. Cook and Weaver conducted a case study of seven rural high school teachers who took part in an NSF-funded STEM project (*Research Goes to School*) to find out how they implemented PBSI units in their classrooms. The teachers developed their project-based science units as part of a summer institute provided by the project. Analysis of their data, which consisted of video recordings, observations, and semi-structured interviews, indicated that the teachers were able to implement their PBSI units with "partial fidelity of ... the instructional features of PBL identified for this study" (p. 31). However, there was variability in implementations and "most of the PBL features were implemented in a less than optimal manner where elements of best practice were present but not consistently implemented" (p. 31). They found that where the teachers encountered the most challenge were in making connections between science content and context of investigations (biofuels) meaningful to the lives of their students, and acting as facilitators as opposed to performing teacher-centered procedures. The limitation of this study, as acknowledged by the researchers, is that they could not account for the prior knowledge and teaching practices of the teachers. Similarly, Park Rogers et al. conducted a case study of three ninth-grade teachers (one math and two science) for one year to determine their first year of experience with PBSI when it was implemented schoolwide. In discussing their findings, they noted:

> Our analysis of teacher orientations has attempted to balance teachers' backgrounds with their experience in implementing a project-based learning curriculum for the first time. Such a shift can be quite challenging, and it is important to attempt to understand the kinds of factors that make it easier or harder for teachers to undergo this change. In examining these three teachers, it is clear that the ideas that one holds about the nature of learning and knowing has significant implications for the kinds of changes to one's practice that one is willing (or able) to make. (p. 911)

Bencze and Bowen, on the other hand, implemented PBSI as part of a university preservice course and found that exposure to this approach to science instruction influenced preservice teachers thinking about science inquiry and decisions to incorporate PBSI features in their future teaching. Tal, Krajcik, and Blumenfeld studied urban teachers who enacted PBSI in their classrooms and noticed a qualitative difference in how they related to their students. They found that the teachers developed positive dispositions about their work despite the challenges they faced and were able to project these dispositions in their own classroom activities and help their students not only to perform better but also focus on their learning. Freeman, Marx, and Cimellaro conducted PBSI institutes for 58 teachers and discovered that the physical and psychological settings for the institutes, the technologies used, and the need to establish a balance to allow teachers to informally interact with each other can either enhance or hinder the institutes.

Engagement of Students in PBSI

According to Gibbs (2014), the term "'student engagement' is now used to refer to so many different things that it is difficult to keep track of what people are actually talking about" (p. 1). In this book, it is essential to provide a working definition so that there is clarity. Student engagement in PBSI refers to students taking responsibility for their own learning throughout the project cycle (actively planning, collaborating with each other, using tools and technology to implement projects, and assessing or evaluating their own learning with their teachers as facilitators and mentors). Student engagement in PBSI was reported in eight studies (Lee, Lee, & Lee; Krajcik; Chin & Chia; Wu &

Krajcik; Moje et al.; Crawford, Krajcik, & Marx; Krajcik et al.). The type of student engagement reported was similar with the exception of one study. For instance, Wu and Krajcik explored students' use of inscriptions (i.e., graphs, photos, diagrams, data tables, symbols, etc.) in PBSI and found that

> when the seventh graders were scaffolded by the teachers and provided with social, conceptual, and material resources, they were able to use various inscriptions to demonstrate meaningful inscriptional practices such as creating and using inscriptions to make arguments, to represent conceptual understandings, and to engage in thoughtful discussions. (p. 869)

Crawford, Krajcik, and Marx concluded,

> Engagement of students in this study as they made decisions and grappled with designing investigations reinforces other studies describing the power of question-driven instruction. ... By the end of the study students demonstrated, to a large extent, an understanding of the various ways to carry out scientific investigations and some of the issues in designing investigations, and collecting and interpreting data. This study highlights the importance of allowing students to choose their own questions to explore, that when coupled with opportunity to exchange ideas and substantial teacher support, enhances the collaborative aspects of the classroom. (p. 721)

Krajcik et al. found that students who participated in a rich PBSI curriculum were very successful with planning and carrying out investigations. However, they struggled with data collection and analysis; therefore, they were lacking in these process skills. Moje et al., on the other hand, found a different type and level of student engagement in PBSI. They noticed that despite its benefits, "project-based pedagogy has the potential for conflict or confusion," particularly when students' first language is not English, which is the dominant language for scientific discourse in most classrooms. The interaction of students' contextual factors (community, language, and culture) and scientific discourse in the classroom poses a real challenge, and how to address it remains an unanswered question.

Impact on Student Learning

Various researchers investigated the impact of PBSI on student science learning outcomes with varying positive results (Han, Capraro, & Capraro; Bilgin, Karakuyu, & Ay; Sola & Ojo; Marx et al., 2004; Schneider et al., 2002). Bilgin, Karakuyu, and Ay and Sola and Ojo used quasi-experimental methods with treatment and control groups, and multiple instruments or measurement scales. In both studies, they found significant differences between the groups, with the PBSI groups performing better than the control groups. Schneider et al. (2002) found that students who participated in a project-based science curriculum outperformed their counterparts on a national measure of science achievement. Using hierarchical linear modeling, diverse sample of students, and longitudinal data set, Han, Capraro, and Capraro investigated whether participation in STEM PBSI impacted student achievement differently. In their concluding statement, they note: "The results, that low achievers and Hispanic students' growth rates were statistically significantly higher through STEM PBLs, should be considered by policy makers, educators, and teachers in designing differentiated instruction" (p. 1110).

Applications of PBSI Within and Across Science Disciplines

Out of the 71 studies identified in this review, 41 focused on the application of PBSI in various science disciplines and grade levels. The disciplines include biology, chemistry, physics, engineering, technology, earth and space science, environmental science, and environmental chemistry. Grade levels covered were from high school to college. Below is summary of findings by disciplines.

Biology

Seven studies reported the use of PBSI to teach biology content. Three of the studies were conducted at the high school level and four at the college level. Duncan and Tseng implemented a project-based genetic curriculum to study the learning progression of high school students in genetics. They organized a collaborative design team of four secondary teachers, including scientists, to design and implement the curriculum. Using pre- and post assessment interviews and video-based classroom observations, they note, "Our findings suggest that the majority of our students did develop an understanding of the central role that proteins play in genetic phenomena and were able to reason about some functions of proteins. However, students' repertoire of protein functions was rather limited even after instruction" (p. 47).

Similarly, Alozie, Eklund, et al. (2010) also studied high school students and how they learn genetics in a PBSI environment. At the conclusion of their study, the researchers identified three challenges and provided suggestions. These included (1) difficulty assessing students' products or artifacts (suggestion: use more formative assessments), (2) students are not socialized to work collaboratively and to engage in open dialogue about their work (suggestion: assign cognitive roles to students, encourage evidence-based discussions), and (3) time constraints (suggestion: teach standards-related material first and have students focus on project work later). Cook, Buck, and Park Rogers conducted a case study to investigate how a ninth-grade science teacher and his students teach and learn evolution. The researchers were interested in exploring the ways in which PBSI supported or hindered cognitive engagement with evolutionary theory. Upon analysis of their data, they found that PBSI allowed student reflections and openness to multiple perspectives and "enhanced procedural types of cognitive engagement" (p. 22). However, the teacher was not able to facilitate cognitive engagement of evolutionary theory and unable to enact a deeper discussion beyond "presenting evolution as a polemic and false dichotomy in science, rather than a social controversy" (p. 22). In an earlier study, Cook proposed teaching evolution by contextualizing and using real-life problems, such as the MARS (methicillin-resistant Staphylococcus aureus) outbreak, with embedded assessment, journal writing, and student reflections as way to develop cognitive engagement on evolution. Brickman et al. used PBSI to help college undergraduates in biology develop scientific literacy skills. They developed and implemented the PAL approach (project-based applied learning), which consists of four steps: introducing a compelling problem, acquiring knowledge and skills, making out-of-class investigations, and performing evaluations. Although the researchers reported positive feedback from their students, some students expressed ambivalence toward taking responsibility for their own learning.

Kosinski-Collins and Gordon-Messer focused on a specific skill that they determined biology students need and that was scientific writing. To accomplish their goal, they had students write the scientific purpose of their laboratories instead of their objectives. They provided feedback and encouraged students to practice and found that students' abilities to write scientific abstracts and overall conceptual understanding of scientific procedures improved. Wright and Boggs planned and implemented a quarter-long cell biology course using PBSI. They organized students into teams of four to five students and had them conduct and present their research projects in a public forum. Comparison of biology grades before and after taking the cell biology course indicated improvements and so were students' dispositions toward science, creativity, and engagement in biology.

In addition, one of the researchers who was also the instructor for the course reported great satisfaction and felt "energized as a teacher and as an individual" (p. 153). Similar to Brickman et al., there were some students who were challenged by the idea of learning on their own with the instructor as facilitator. Regassa and Morrison-Shetlar (2009) implemented a PBSI in a molecular biology course over four semesters and collected multiple levels of data. Their analysis shows that "project-based learning increased students' knowledge base, confidence levels, critical thinking, and analytical skills. By the end of the course, indirect and direct measures indicated that the majority of students gained the confidence and knowledge needed to problem-solve independently using molecular biology approaches" (p. 66).

Chemistry

Out of the 11 articles dealing with the implementation of PBSI in chemistry classrooms or labs, 5 were classified as EVRA (de los Santos et al.; Tsaparlis & Gorezi; Adami; Draper; Juhl, Yearsley, & Silva). Two articles were classified as ARA (Hike & Beck-Winchatz; Short, Lundsgaard, & Krajcik), two as ERA (Sola & Ojo; Barak & Dori), and two as TRA (Kiefer et al.; Wilson, Parkin, & Thomas). Three of the articles on PBSI in chemistry described the participation of high school or secondary students, while six articles focused on undergraduate students' participation.

Assessment of student learning in project-based chemistry classrooms at the secondary level showed significant gains in "point charges and Coulomb's law, instantaneous dipoles and London dispersion forces, and comparisons between the various types of intermolecular forces" (Short, Lundsgaard, & Krajcik, p. 43). However, according to the researchers, students were not able to develop a deeper understanding of some of the concepts, such as "dipoles/polar bonds and differential pressure" (p. 43). Similarly, Sola and Ojo also found significant gains in students' understanding of experimentation on separation of mixtures in project-based chemistry labs. One article reports that secondary students who implemented a unit on project-based chemistry took ownership of their own learning. They integrated literacy practices such as group discussion, writing, and reflections and were able to meet the Common Core Standards. In addition, students evaluated their hypotheses, proposed alternative explanations, and revised experimental designs (Hike & Beck-Winchatz).

Undergraduate PBSI chemistry covered topics such as sol-gel applications in stones, organic chemistry, crystallography, physical chemistry, environmental chemistry, and analytical chemistry (see table 2.1). Overall, the articles described the implementation of a chemistry lab, unit, or course using a project-based approach. The collective findings from these articles suggested that PBSI was successfully implemented in chemistry classrooms and labs and students developed both science content and process skills. For instance, Juhl, Yearsley, and Silva note that implemented PBSI

> provided a unique mechanism to introduce and review important technical skills, communication skills, computer skills, and interpersonal skills necessary for employment. The interdisciplinary nature and extensive scope of this project gave students a sense of the responsibilities, independence, and self motivation that will be necessary to succeed in their future careers. Difficulties such as sample contamination and calibration problems provided students with a valuable taste of the true scientific experience. ... Our experience suggests that an extensive, interdisciplinary project offers a challenging and meaningful alternative for delivering critical science skills and experiences to students. (p. 1433)

A similar experience was encountered by Adami, who summed it up this way,

> The project increases the self-motivation of students and gives them a sense of responsibility and independence, stimulating their interest in science. The learning-lab activity, based on a project, introduces important technical instrument skills, computer skills, and communication skills by means of interactions with local environmental agencies. Students also have the opportunity to present, often for the first time in their undergraduate career, their scientific results in a public forum. (p. 256)

One of the key principles of PBSI is that it should be prepared with the right scientific dispositions so that students can take responsibility for their own learning. This principle is repeated across the articles and reiterated by Tsaparlis and Gorezi, in the following words: "The originality of the projects and the feeling of ownership and responsibility contributed to the dedication and enthusiasm of the students during the performance of the experiments" (p. 669). The effects of technology enhanced project-based chemistry on undergraduates' science learning was investigated by Barak and Dori using a quasi-experimental method with a treatment and control group. The researchers found:

> Students who participated in the IT-enhanced PBL performed significantly better than their control classmates not only on their post-test but also on their course final examination ... that the construction of computerized models and Web-based inquiry activities helped promote students' ability of mentally traversing the four levels of chemistry understanding: symbolic, macroscopic, microscopic, and process. More generally, ...

that incorporating IT-rich PBL into freshmen courses can enhance students' understanding of chemical concepts, theories, and molecular structures. (p. 118)

It is important to note that implementation of project-based chemistry is not without challenges. Kiefer et al. report that when students were encouraged to implement project-based chemistry labs they ended up creating hazardous substances by accident, and that is one of the risks instructors must consider when implementing PBSI; therefore, they should exercise "constant vigilance, careful review of student proposals for hidden hazards, and prudent safety oversight" (p. 686).

Physics

The project-based physics articles were distributed as follows: two were EVRA (Liu; Pearce), two were ARA (Toolin & Watson; Wilhelm & Confrey) and two were ERA (Wilhelm, Thacker, & Wilhelm; Milner-Bolotin & Svinicki). An examination of the articles reveals that a group of high school students learned about waves and sound using PBSI. Pre- and posttests showed that the students not only made significant gains in waves and trigonometry knowledge, but they also developed understanding of concept superposition, improved understanding of amplitude and period, and conceptual understanding of physics (Wilhelm & Confrey). Toolin and Watson also worked with high school students to implement project-based physics. They had students come up with their driving research questions and designed sustainable energy projects to answer their questions. Then they presented their products-artifacts to their parents and other community members. Through this effort, students were able to increase awareness of sustainable energy projects in their own community and also learned to connect physics with real-world applications.

Wilhelm, Thacker, and Wilhelm investigated the effects of an introductory project-based physics curriculum on students' conceptual understanding of physics. Participants came from a diverse group of undergraduate students, including preservice teacher candidates. Student projects focused on investigating the relationship between mathematical and physical characteristics of slope within the context of a velocity and time graph as well as energy conservation and transfer. When they were assessed using a Force Concept Inventory, the results show that the students achieved a higher normalizing gain score than students who were taught physics using a traditional approach. Milner-Bolotin and Svinicki implemented a project-based physics course for nonscience majors and found that by having students pose their own questions, nonscience majors were able to develop the capacity and motivation to learn science. However, their study also raised several interesting, but unanswered, questions including the key factors in PBSI that are responsible for cognitive and attitudinal outcomes. Pearce notes four noticeable changes from his students after implementing a project-based physics course such as "(1) observed increase in student motivation to learn physics, (2) observed time invested by students on projects outside of structured class, (3) student comments on end-of-semester instructor evaluation forms, and (4) a modest increase in pre/post-exam grades" (p. 166). In addition, he adds, "It is interesting to note that the largest improvement was observed in the academically poorest class, whose class average increased by 12% (from a low D to a C), while the highest achieving class showed no improvement. These preliminary results indicate that these projects are most beneficial to those students who are having difficulty learning physics" (p. 166).

Earth Science/Environmental Science

There were five articles that focused on earth and/or environmental science content (Colaianne; Vacchina & Aguirre; Colley, 2006; Tal & Argaman; Berenfeld). With the exception of Tal and Argaman, all the articles focused on student investigations of environmental questions or problems using a project-based approach enhanced with tools and technology. Colaianne's article demonstrates how to use the Intergovernmental Panel on Climate Change (IPCC) framework to teach the cross-cutting concepts in the NGSS, while Berenfeld describes the Global Lab Project (a PBSI curriculum for middle and secondary schools implemented in the United States and abroad), its objectives, implementation strategies, tools, technologies, and student outcomes. Colley (2006) reports on the planning and implementing of a project-based park ecology unit for inner-city high school students in an afterschool program. Analysis of student project reports and artifacts indicate that students were able to pose driving research questions, plan, and collect data on their projects. However, they lacked the knowledge and skills in conducting meaningful data analysis. In addition, Colley (2006) finds that planning and managing resources (tools, materials, technology) in a resource-poor learning environment was a challenge.

Instead of high school students, Tal and Argaman focus on environmental science teachers and how they mentor students in a PBSI learning environment. They wanted to investigate the type of mentoring skills needed by experienced and inexperienced teachers, stages of the projects that are mentoring intensive, and teachers' perceptions of challenges with mentoring students on their projects. Using an interpretive research approach (grounded theory, journals, questionnaires, interviews, and observations), Tal and Argaman find that experienced teachers identified more skills required for mentoring students' inquiry than inexperienced teachers. In addition, experienced teachers demonstrated more open-ended and flexible mentoring patterns, while inexperienced teachers were more structured and authoritative in their mentoring patterns. The Vacchina and Aguirre article is discussed in chapter 5. Please refer to that chapter for details.

Engineering

Four articles covered topics related to engineering (Shen et al.; Palmer & Hall; Doppelt; Mills & Treagust). Of the four articles, three were classified as EVRA and one as ERA. All but one article focuses on undergraduate engineering students. Shen et al. implement a sustainable design course using a project-based approach enhanced with technology (building analysis software). Results indicate that "Building Information Modeling (BIM), coupled with building energy analysis software and used with PBL, provided a good pedagogical as well as a suitable technical platform for teaching sustainability in a building design and construction class. The majority of students found BIM-PBL to be an effective knowledge enhancement tool" (p. 146). However, the researchers warn, "Due to the pilot nature and small number of students involved in this study, these conclusions need to be confirmed by larger-scale studies" (p. 146). Palmer and Hall find no significant difference across demographic groups in a PBSI unit designed to introduce college students to engineering design and professional practice. They propose that the number of design projects be reduced from three to two per semester and better methods of assigning individual grades (scores) within group work be developed. Doppelt looks at the intersections of engineering education, creative thinking in the design process, and the assessment of project-based learning. The researcher used a qualitative research method to investigate how students design projects, the roles their teachers play in the process, and how their learning was assessed. Findings from this study demonstrate

> Pupils in high school can create, design, implement, control, and document authentic, real-life projects instead of solving well defined problems prescribed by the teacher. ... Furthermore, pupils have proven through their projects that they are

capable of dealing with the "large definition of DESIGN"—that the DESIGN activity does, in fact, encompass the entire process of planning, designing, constructing, and managing the development of a product. The CTS (creative thinking scale) has enabled teachers and researchers to set goals for the pupils (and for the teachers) during the PBL. The consistency of the judges' scores and the successful application of the criteria developed by the teachers strengthened their validity. The findings of the assessment process indicate that the CDP (creative design process) and the CTS are useful and can be implemented by teachers who have participated in a suitable in-service training. (p. 12)

Mills and Treagust examine the application of PBSI and problem-based learning in engineering education. They focus their study on examples where both approaches have been used and determine their relevance and overall effectiveness. In their conclusion, they note, "It has also been demonstrated that the engineering profession and academics are more familiar with the concepts of projects in their professional practice, than with the concepts of problem-based learning. It therefore seems that project-based learning is likely to be more readily adopted and adapted by university engineering programs than problem-based learning" (p. 13).

Chapter Summary

The purpose of this chapter was to review the body of research on PBSI with particular emphasis on peer-reviewed or refereed studies conducted between 1990–2015. Out of more than 300 documents identified, 71 met the criteria for inclusion in this review. The articles used for the review are shown in table 2.1 and are organized in chronological order, beginning with the most recent articles published in the current year and ending with the earlier or oldest ones published in the early 1990s. Based on the 71 studies, approximately 16,936 students and 704 teachers were reported to have taken part in PBSI research-related activities, although these numbers may not represent the true population of teachers and students because not all articles reported this piece of information. To organize the review, the articles were classified into four groups (ERA, EVRA, TRA, and ARA), based on method of study. Five themes emerge from the review and include the following: PBSI curriculum materials, engagement of teachers on PBSI professional development, engagement of students in project-based science learning, impact on student learning, and applications of PBSI within and across science disciplines. Each theme was discussed with reference to the relevant articles. In general, the review demonstrated a viable body of work that tends to suggest that teachers and students from K–16 education levels are implementing PBSI and, in most cases, these have been successful. There are, however, still challenges ahead and unanswered questions that require further investigations.

Food for Thought

The following questions will help the reader reflect on the chapter:

1. What does the research say about developing and evaluating PBSI curriculum materials?
2. What does the research say about engagement of teachers on PBSI professional development?
3. What does the research say about engagement of students in project-based science learning?
4. What does the research say about impact on student learning?
5. What does the research say about applications of PBSI within and across science disciplines? (Choose any two disciplines.)
6. The articles in this review are classified into four groups. Which group is most common and which group is least common?
7. What types of PBSI research articles do you think are absent and/or needed the most?
8. What PBSI-related research question would you like to see investigated? Briefly describe your proposed method of study.

· 3 ·

HOW TO PLAN FOR PROJECT-BASED SCIENCE INSTRUCTION

Chapter Overview

In order for teachers and their students to plan for PBSI, they must demonstrate knowledge and understanding of the underlying principles or concepts of this instructional approach to science education. In addition, they must know what the various stages of PBSI are and the roles, responsibilities, expectations, and dispositions of teachers and students at each stage of the PBSI process. In this chapter, I will identify and describe the basic principles or concepts of PBSI. Then I will describe the main stages involved in PBSI. The main stages of PBSI include: (1) an orientation process, (2) identifying and defining a project, (3) planning, implementing, documenting, and reporting project findings, and (4) evaluating project learning, taking action, and/or proposing new projects. Collectively, these stages are referred to as the *project cycle*. This chapter will focus on the first three stages of the project cycle (i.e., orientation process, identifying and defining a project, and project planning). The other stages will be discussed in detail in chapter 4: "Implementing Project-based Science Instruction," and chapter 6: "Evaluating Project-based Science Instruction."

Underlying Principles and Concepts of PBSI

Science teachers and their students who want to implement PBSI in their classroom should understand the following basic principles. In a PBSI learning environment,

1. The roles of teachers are to facilitate, advise, guide, monitor, and mentor their students, not only to conduct lectures and laboratory work.
2. Students are responsible and accountable for their own learning. Their roles are to be active learners, co-investigators, and teammates who contribute to the learning process.
3. The classroom is a dynamic learning environment where roles constantly change. For instance, in some cases, teachers will become students and students will become teachers.
4. Lesson planning is not only about the method of delivering and assessing science content. It is also about defining the area of study, identifying the learning environment and process, selecting the resources and time required, identifying possible learning challenges, and selecting the appropriate formative and summative methods of assessing learning outcomes.
5. Science learning is based on what is relevant to students' lives and communities, not just on textbooks, curriculum guides, and content standards.
6. There are tangible outcomes, products, and/or student artifacts at the conclusion of PBSI.
7. The applications of science and engineering through the design, use, and/or calibration of scientific instruments, tools, and technologies are encouraged and promoted.
8. PBSI requires extensive preparatory work on the part of teachers, although this is usually at the beginning. When students get used to taking responsibility for their own learning, there is ample time for teachers to work on related tasks such as researching for grants, identifying and selecting resources and new topics, and developing supporting materials and assessments, among others.

It is important to note that Principles 1 and 2 are not mutually exclusive but go together as identical twins. By teachers playing the role of facilitators, advisors and mentors, while students take responsibility for their own learning, does not mean that

> teachers abdicate their responsibilities. Like adults, children [or students] can be responsible only for what is within their capacity and control, and teachers must retain responsibility for determining the areas within which children's decisions are desirable and effective. Unless the adults with whom they learn set limits to these areas, children [or students] … are likely to feel insecure. In order to experiment confidently and usefully with making their own decisions about what to learn, how to learn and when to learn, they need to know that in the background are adults who are willing to help but ready to point out considerations children [or students] may not be aware themselves. (Muir, 1970, p. 32)

A Day in the Life of a Project

The first step in the PBSI process is orientation. Orientation is a process whereby the science teachers and their students spend some quality time discussing expectations, requirements, roles, and responsibilities prior to implementing projects in the science classroom. It should not be assumed that students will be interested and/or actively participate in PBSI. The point to remember is that most students are not socialized to take responsibility for their learning and therefore need some guidance at the beginning of PBSI. The next step following orientation is identifying and defining a project. This is followed by planning a project, which is a process of thinking, reflecting, discussing, and documenting the driving question to be investigated. Once project plans are thoroughly vetted by teachers, then students can be allowed to implement them. Implementing a project is a very hands-on, practical matter. It is a process of "doing science." During the implementation stage, students are expected to work collaboratively in a spirit of cooperation and mutual respect. The next stage in the process is known as documenting and reporting project findings. During this step, students are expected to prepare written reports and present them for peer review.

After documenting and reporting their projects, students reflect on their learning prior to and after conducting their projects in terms of science content, process skills, and dispositions. Depending on context, a performance-based assessment is administered to determine knowledge and understanding gained from project work. This is the evaluating and assessment of learning stage. Sometimes PBSI terminates at this stage. Other times, students are encouraged to put into practice what they have learned from their project through advocacy, by interaction with local authorities to bring about desirable change in their school community. This is the taking action stage. Figure 3.1 graphically represents the different stages of the project cycle. There are seven main stages in the cycle, although each stage could be broken down into sub-stages. The figure is designed to be read in a clockwise direction and the stages act as building blocks, which are all interconnected and supportive of each other as indicated by the two dotted cycles.

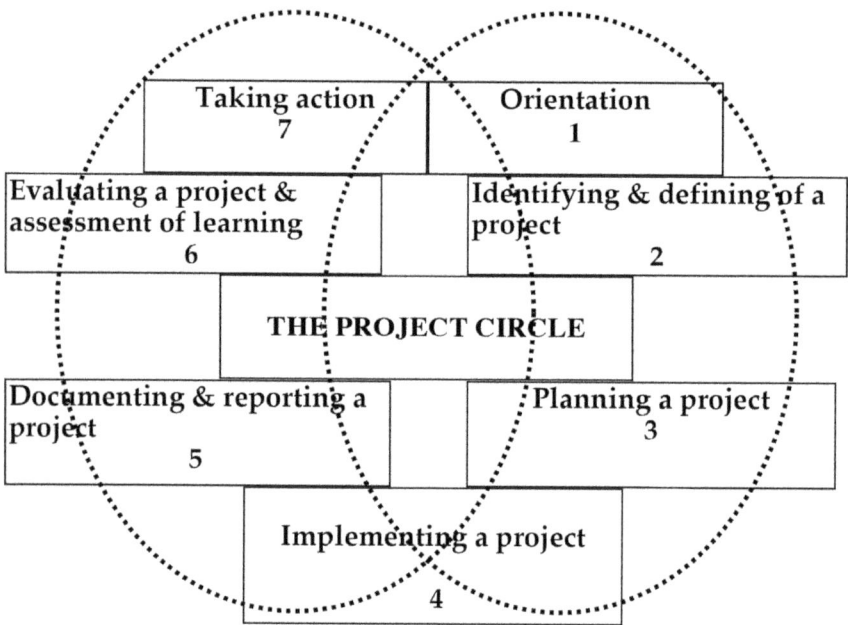

Figure 3.1. The Project Cycle.

Orientation to PBSI

As mentioned earlier, orientation is the number one order of business in PBSI. During orientation, students should be informed that there are some basic expectations of PBSI, the most important of which is taking responsibility for their own learning.

Students should be instructed on the importance of collaboration in science. For instance, they should learn how to work together in group settings, how to interact and relate to each other, what group dynamics is, the benefits of sharing project work, and how to deal with the challenges inherent in collaborative projects. They should learn that before conducting any project work, safety issues, procedures, and precaution must be observed. Assessment of learning in a project-based learning context is very different from assessment in a non-project-based learning context. In a project-based learning context, because the learning is active, practical, and collaborative, assessment of learning should mirror how students learn, hence assessment should be performance based. Students should therefore be instructed and exposed to the different types of performance-based assessment (e.g., open-ended performance task, close-ended performance task, portfolios, presentations, essays and reports).

Expectations and Requirements of PBSI

The expectations and requirements for PBSI may vary. However, the following excerpt from Beveridge (1957) illustrates the point well.

> The most successful scientists are ... disciplined by objective judgment of their results and by the need to meet criticism from others. Love of science is likely to be accompanied by scientific taste and also is necessary to enable one to persist in the face of frustration.
>
> Willingness to work hard and tenacity of purpose are further requisites for success in research, as in nearly all walks of life. The scientist also needs some imagination so that he [she] can picture in his [her] mind how processes work, how things take place that cannot be observed and conjure up hypotheses. ...
>
> A spirit of indomitable perseverance has characterized nearly all successful scientists, for most worthwhile achievements required persistence and courage in face of repeated frustration. (p. 187)

Although the above expectations and requirements proposed by Beveridge are not specifically directed toward science teachers and students, they are nevertheless important considerations for anyone learning science or aspiring to become a scientist. The message from the above excerpt for teachers and students planning to implement PBSI is that learning science through projects requires one to be a critical thinker, have a deep love for scientific knowledge, and have a willingness to work hard. In addition, it requires teachers and students to show imagination, which is essential for conceptualizing ideas, formulating hypotheses, and/or posing interesting questions. Students and teachers must demonstrate a willingness to persevere and persist in the "face of repeated frustration." All of the above qualities or characteristics may seem obvious, but they need to be reiterated during the orientation phase of PBSI.

Another requirement for PBSI that both teachers and students need to be aware of is that "in reporting an investigation, the author is under obligation to give due credit to previous work he [she] has drawn upon and to anyone who has assisted materially in the investigations" (Beveridge, p. 194).

Roles and Responsibilities

Project-based science instruction and learning entail some amount of responsibility on the part of both teachers and students. The responsibilities of a teacher in a project-based science classroom is best captured in the following words:

> My role in cultivating such an atmosphere is critical. I'm constantly moving from one student to the next—watching, listening, asking questions, challenging, offering suggestions or lending a hand. I am much less of a teacher than a facilitator, guide, and resource. (Wolk, 1994, p. 44)

> In the project method the role of the teacher is a varied one. He [she] is a leader, chairperson, chief interlocutor, coach, umpire, taskmaster, authority, judge, adviser, sympathetic listener, chief performer, examiner, guide, or friend as occasion may require. (Hosic & Chase, 1924, p. 28)

The above quotes indicate that under project-based science instruction, teachers will have to reinvent themselves. They will be expected to take on more proactive and dynamic roles. They will be learners as well as teachers. They will be responsible for organizing the learning environment, providing advice on project planning, and locating resources. They will also mentor students throughout the implementation process and help them assess their own learning as well as the outcomes of their projects. This may seem to be a large obligation on the part of the teacher, but it is not. The students also share in the responsibility. For instance, before students start their projects, they must develop their own plans, which must include questions to be investigated; methods, materials, and technologies to be used; implementation plans; and methods of gathering, analyzing, interpreting, and reporting project outcomes. In addition, students must discuss with their teacher how their work will be assessed to determine their learning, as well as present their final product to their peers. This shared responsibility in teaching and learning makes project-based science instruction very democratic and exciting to pursue.

The ability and willingness to collaborate with others on project work is also a major requirement for project-based science instruction. Students must be made aware of the importance of collaboration in conducting project work. To ensure this, teachers should spend some time discussing group dynamics. They must emphasize that collaboration is a prerequisite for success in scientific investigations. This is because most scientific investigations require collaboration in tasks such as design and testing of instruments, data collection and sharing, data analysis, and peer review of completed work. In addition, it is generally believed that having more "heads" working on a question or problem is better than having one "head." Also when scientists collaborate, they complement each other and are able to generate valid and reliable scientific knowledge.

The discussion of collaboration must also address the advantages as well as the disadvantages of working in groups. However, students should be made aware that the advantages of working in groups outweigh the disadvantages (Jones & Carter, 1998). It is important for students to understand that working in groups implies that they will have to support each other instead of competing with each other, complement each others' weaknesses with each others' strengths, offer constructive criticism as opposed to destructive criticism, make collective decisions as opposed to individual decisions, and take collective responsibility as opposed to individual responsibility.

There are several strategies for setting up groups. One strategy is to divide the class into groups of three or four. Smaller groups tend to work better compared to larger groups. Where possible, the teacher must make sure that each group is balanced in terms of interests, skills, academic ability, and background status. Students must be aware that they will be required to work as a group throughout the lifecycle of their projects. They will be expected to make sacrifices and commit to the group. When disagreement or conflict arises, it must be resolved through dialogue, a convincing argument, and evidence. Students must be told that they will be held accountable for their own learning as a group at the end of the day.

How to Identify and Define a Project

The world we live in is full of interesting problems and challenges, such as how to slow down global warming; produce cleaner and renewable sources of energy; produce food without the use of synthetic chemicals and hormones; recycle garbage and dispose of hazardous waste; conserve water, soil, and air quality; increase access to information technology; find a cure for HIV-AIDS; prevent pandemics such as Ebola; protect endangered species; and develop environmentally friendly cars, to name a few.

One way to identify a project is to have students think of an area that is full of interesting problems or challenges and also of interest to them. Then have students write down on a piece of paper one project that they would like to undertake. After writing their favorite project down, students must also answer the following questions: (1) Does our project have a clear purpose? (2) Can we investigate our project in the amount of time allocated? (3) Do we have the resources to investigate this project or can we investigate this project with the available resources? (4) Will our project benefit other people? (5) Will our project contribute new knowledge to the field? (6) Is our project timely? If students answer yes to all six questions, then perhaps the group has a potential project. If, on the other hand, any of the answers are no, then the group may want to consider another project, because chances are that it is going to be a difficult project to investigate.

Another way to identify a project is to have students visit a library and go through the section that houses the indexes and abstracts of various publications. By going through indexes and abstracts, students will be able to see what sorts of questions or topics people are studying. They will also be able to identify and review journal articles that pertain to their topics, or questions to learn more about methods, materials, results, and suggestions for further research. Based on this experience, students can then generate their own questions or repose old questions in new ways.

Alternately, students can use an online search engine on the Internet such as Google, Yahoo, Bing, or similar search engine and type a word or phrase relating to their topic of interest. The results they get will determine the next step in their search. For instance, if their search yields several documents and/or links, they can identify the document or link most appropriate to their needs and conduct further searches until they arrive at what they want. On the other hand, if their search does not yield meaningful results, then they may have to repeat the process. It is important for students to know that conducting an Internet search takes time and requires patience. In addition, it assumes some knowledge and skills in the use of a computer and familiarity with Internet search engines. If none of the above strategies work, students can ask their teacher or an expert on the topic of interest to suggest a potential topic for a project. Once a topic is identified, then the next step is to formulate a project question. The project question must be framed with "question identifiers," such as what, where, when, why, who, and how. It is important to make the question as specific as possible. If a project question is too broad, then it becomes difficult to investigate. For example:

Example 1

> **Broad Project Question:** How healthy is the ecosystem at Brookville Park in the city of New York?
>
> **Specific Research Question:** How does the water quality in the Conselyeas Pond at Brookville Park in the Borough of Queens, New York City, compare with New York State environmental water quality standards?

As can be seen in Example 1, the specific project question refines and narrows the broad project question down to (1) water quality, (2) in the Conselyeas Pond, (3) at Brookville Park in the Borough of Queens, and (4) compared with New York State environmental water quality standards. In addition, it helps the investigator to focus on the substance of the project rather than the generalities. One rule of thumb to make project questions more specific is to have students examine the key terms or words in their questions and ask themselves what they mean. Then try and clarify the terms. The students should assume that their readers are not familiar with their project questions and it is, therefore, their responsibility to make their questions as clear as possible. Also they must consider the context in which their questions are being asked. According to Cook, Buck, and Park Rogers (2012), "Construction of the driving question in a PBL is critical to students' motivation and engagement; the question should not be so constraining as to predetermine the project's outcomes, nor should it be so broad that it would overwhelm and de-motivate students' attempts to learn and engage in problem-solving" (p. 25). Table 3.1 provides a tool for testing a research question.

The same procedure described above for formulating a project question could be used for formulating a hypothesis. However, a hypothesis differs from a question because it is framed in the form of a statement that is testable. It is a guess expressed in formal language. A hypothesis usually implies that the investigator is going to conduct an investigation, collect data, and test or verify to see if the data fits the hypothesis. A hypothesis contains variable(s) that must be defined. The word *variable* refers to a word or thing that changes in meaning according to its context. According to Isaac and Michael (1979),

Table 3.1. Framework for Aligning a Project Question to a Science Learning Standard

Project Questions	Knowledge or Concepts	Skills	Dispositions (Attitudes)
Learning Standards	Knowledge or Concepts	Skills	Dispositions (Attitudes)
Alignment Score: 0 = Not aligned 0.5 = Partially aligned 1 = Fully aligned			

there are three types of variables: dependent, independent, and control. The dependent variable is what you are interested in studying. It is your outcome. The independent variable is the factor that you change or manipulate in an experiment or investigation. It is your treatment. For example:

Example 2

> **a. Hypothesis:** The growth rate of a corn plant is related to its exposure to sunlight.
>
> *Growth rate* would become your **dependent variable** because it is your *outcome*, and *sunlight* would be your **independent variable** because it is the *treatment*.
>
> **b. Hypothesis:** Smoking cigarettes causes lung cancer.
> **Independent variable:** smoking cigarettes
> **Dependent variable:** lung cancer

A control variable is a variable that does not change. It is what is held constant in an experiment or investigation. For example:

Example 3

> **Hypothesis:** Malarial epidemics are related to climatic factors, controlling for human activity.
>
> *Malarial epidemics* would become your **dependent variable** because it is your *outcome*, and *climatic factors* would be your **independent variable** because it is the *treatment or what would change*. The **control variable** is human activity because it will not change in the investigation.

The decision to formulate a question or a hypothesis will depend on the type of topics students select, the method of study to be used, the resources available, and the time required to implement the study. For instance, social science topics or topics relating to human behavior could best be studied using questions, while natural and physical science topics could best be investigated using questions and/or hypotheses. Because it is more challenging to observe variables relating to human behavior, and to construct valid and reliable instruments to measure them, it is a lot harder to test hypotheses relating to them (Jaeger, 1993; Wilson, 1990).

After identifying their driving research questions and defining them, it is important for students to review previous work done on their questions or topics by other people. This is necessary because it helps students identify research methods, materials, and technologies that may be appropriate for their projects. Students could avoid reinventing the wheel and repeating other people's mistakes. In reviewing previous work done, it is important to focus on the following sections of a publication: (1) research methods or procedures, (2) data collection instruments or procedures, (3) results, (4) discussion/findings, and (5) references. The research methods or procedures section provide information on research methods or procedures, experimental design used, sampling techniques, instruments, data collection techniques, data analysis tools and procedures, materials, equipment, and technologies used to conduct the research. In the results and discussion-findings sections, you will find hard data from the study, whether or not it was conclusive, and specific suggestions about future research. The reference section provides a list of sources that you could refer to in order to learn more about the research question. It contains names of authors, titles of publications, names of publications, years of publications, and page numbers. Students may even find addresses of the author(s) in some references, which could be helpful if they want to contact or write to the author(s).

Aligning Students' Project Questions to Science Learning Standards

According to the National Research Council (1996), "Inquiry into authentic questions generated from students' experiences is the central strategy for science teaching" (p. 31). PBSI, as a method of science instruction, is consistent with inquiry-based science instruction because it allows students to formulate questions of interest and investigate them within a particular context, using appropriate resources and teacher and/or expert guidance for the purpose of acquiring science content knowledge and process skills.

However, it is important to note that under PBSI, students may select project questions that do not align with state or national science learning standards. It is, therefore, the responsibility of the teacher to help students align their project questions to learning standards. Science learning standards mean different things to different people. In addition, they carry with them some basic assumptions about students, teachers, instruction, learning, and the environment in which science learning takes place. The teacher must be knowledgeable and skilled in analyzing and interpreting the learning standards. Only then will they be able to help their students formulate questions that are aligned with state and national science learning standards (NSES and NGSS).

To analyze and interpret science learning standards, teachers must have a framework or tool to conduct their analysis. Figure 3.2 shows an example of a framework used to analyze science learning standards (Colley, 1998). Although the example is taken from the New York State science learning standards, it could be applied to other standards. It presents one of New York State's science learning standards and six questions are used to examine it. Note that only the minimum responses are provided. The main assumption behind this framework is that teachers are more likely to implement a standard if they know the meaning, assumptions, content, and pedagogical requirements of the particular standard. Teachers could use all or modify the questions to review each learning standard before they implement them in their classrooms.

> **New York State Learning Standard for Mathematics, Science and Technology (Standard No. 4)**
>
> "Students will understand and apply scientific concepts, principles, and theories pertaining to the physical setting and living environment and recognize the historical development of ideas in science." Source: University of the State of New York, 1996
>
> **Application of Framework**
>
> *What is the meaning of the standard?*
> This standard means that teachers must prepare students to understand and apply scientific concepts in earth science (physical setting) and biology (living environment). For example, using the concept of osmosis and diffusion to explain how molecules move in living cells. Students must also understand how ideas in science evolved or developed. For example, how Einstein's theory of relativity or Darwin's theory of evolution developed.
>
> *What are the underlying assumptions? (State the minimum assumptions)*
> Students have basic language proficiency and can communicate their ideas clearly in writing. Teachers are well prepared in earth science and biology content and know how to apply this content knowledge. Basic resources such as textbooks, teaching materials, laboratory facilities, equipment and supplies will be available.
>
> *What background knowledge and skill does a teacher need to teach the standard? (State the minimum knowledge and two skills)*
> Earth science and biology content knowledge at or above the undergraduate level. Understanding of the historical development of important scientific ideas in earth science and biology. Knowledge and skills in implementing activities that help students understand and able to apply scientific concepts.
>
> *What background knowledge and skill does a student need to learn the standard? (State the minimum knowledge and two skills)*
> Students must have basic reading and writing competency. In addition, students must have basic mathematics and laboratory skills.
>
> *What teaching method(s) or strategies can a teacher use to teach the standard? (State the minimum teaching methods/strategies)*
> Discussion and demonstration of basic concepts
> Project-method to allow students to investigate scientific ideas or theories.
>
> *What assessment method(s) or strategies can a teacher use to assess the learning of this standard? (State minimum assessment methods or strategies)*
> Think-aloud problems, focus-group interviews, project reports and short-response essays

Figure 3.2. Application of an Analytical Framework to Standard 4 of the New York State Learning Standard for Mathematics, Science, and Technology.

Table 3.1 shows a framework for aligning project-driving questions to science learning standards. Such a framework could be used by teachers to evaluate students' project questions prior to their approval. In a project-based science classroom, students must develop and submit project plans to their teachers,

PROJECT PLAN
STUDENT NAMES: PROJECT TITLE: DATE:

1. **Project Question:** What are the physical, chemical, and biological properties of the water at Goodville Lake?
 	
2. **Purpose and importance of project question:** We want to research this question because we want to gain knowledge about the physical, chemical, and biological properties of the water. Living things cannot live without water, and we hope our project will bring more awareness of the importance of water and the need to protect it from pollution.
 	
3. **Methods (list, step by step, how your project will be carried out)**
   a. First the samples will be collected (samples from different parts of lake).
   b. Samples will be tested for physical, chemical, and biological properties.
   c. Physical, chemical, and biological observations will be taken as water testing is going on.
   d. Different reactions during testing will be recorded.
   e. Results from testing will be recorded and proper documents written.
 	
4. **List the tools and materials required**
   a. Water samples to be tested
   b. Water testing kit
   c. Gloves to prevent contamination during testing
 	
5. **Time table (when will the project start and end? How long will it take? days? weeks? or months?)**
The research is expected to take 3–4 weeks.
 	
6. **Identification of roles and responsibilities (who will do what? when? and how?)**
   1. K and J will collect water samples and record biological observations.
   2. G and B will carry water testing kit and conduct testing of water samples for different parameters.
   3. A and M will write down the different physical and chemical observations, compose final report.
   4. All group members will read report for mistakes/typos and take part in presentation.
 	
7. **How will your project be assessed or evaluated?**
Each group member will keep a journal.
Each group member will write a section of the final document.
Each group member will take part in presenting the final results.

Figure 3.3. An Example of a Project Plan.

who are responsible for reviewing their plans to see if they are doable as well as meeting curricular goals and objectives. Teachers can use this opportunity to check students' project questions to see whether or not they are aligned to specific learning standards. Table 3.2 or a modified version could be used by the teacher to review project plans or it could be provided to students to evaluate their own questions in terms of the concepts, skills, and dispositions they will learn from pursuing their project plans.

Another strategy for aligning students' project questions to science learning standards is for teachers to anticipate the type of questions and topics students might choose to conduct their projects and develop a concept map showing all learning possibilities (Helm & Katz, 2001). According to Beneke (2003), "Adding benchmarks or standards to the web [concept map] helps the teacher anticipate what goals are likely to be met through the course of the project and what learning experiences he or she must plan to provide in addition to the project" (p. 81). This means that the teacher is going to make sure that the key elements or requirements of the standards are integrated into or included in the concept map. The teacher can then use the map to guide students in selecting questions for their projects.

Table 3.2. Tool for Testing a Driving Research Question.

CRITERIA	YES	NO
1. Did the research question begin with an identifier such as **why? what? when? who? where? how? did?** etc.		
2. Is there a **question mark** at the end of the research question?		
3. Are all the terms used clear to the reader? Is the research question free from ambiguity?		
4. Is the research question specific?		
5. Did the research question identify the sample, participant, or subject to be studied?		
6. Did the research question include a dependent variable(s)?		
7. Did the research question include an independent variable(s)?		
8. Can the research question be investigated in the short term?		
9. Can the research question be investigated with limited resources?		
10. Is the research question relevant to society?		
11. Is the research question timely?		
12. Is the research question new?		
A good research question should respond "YES" to all the above questions.		

Weizman, Schwartz, and Fortus (2008) propose using what they call the "driving question board" to help students create a visual organizer of their driving research questions. The "driving question board" is "a large poster board that presents the driving question, which is surrounded by sub-questions that are the foci of different section in the PBS unit" (p. 34). According to the researchers, the driving question board provides a tool that students can use to map out their research questions and show how it is connected to the unit, PBSI context, and related learning standards. It "helps students break big questions into smaller ones, and gives them a sense of ownership" (p. 37).

Developing a Project Plan

A project plan is a document that describes the purpose, importance, methods, resources, and activities that a student or group of students will engage in to answer research questions or solve problems. It serves two main functions: (1) to propose what and how students are going to learn and (2) to provide a framework for evaluating what and how the students have learned. There is no single way to prepare a project plan. However, every project plan should include the following: title, driving research question(s), purpose, justification or importance, methods or procedures for investigation, materials, tools and/or technology, and evaluation activities. Figure 3.2 shows an example of a project plan from a group of high school students who implemented a water quality investigation project. The plan has been modified for this book. The complexity and level of detail of students' project plans will vary depending on topic, grade level, ability level, subject matter, curricula goals, instructional time, and resources available.

It is important to emphasize that carefully thought out project plans with well-defined, specific questions, that are anchored in curriculum or unit content, are essential for success in PBSI environments. Sometimes, this stage of PBSI may take more time than expected, but it is worth the effort to ensure that students participate fully in the process and that they feel they have some ownership of the process. There is no point implementing a project that is going to fail or yield meaningless outcomes. According to Dickinson and Jackson (2008),

> We have found that students respond best to a purposeful transition from structured to guided to open inquiry. ... Projects should start small, working from mini-projects to larger scale projects. Small projects are easier to manage and fit into tightly packed

curricula. Piloting a smaller version allows the teacher to identify and fix major problems before committing to a longer time frame and greater investment of resources. (p. 31)

Table 3.3 represents an example of a project timeline. A project timeline is a graphical or visual representation of project activities and proposed dates of implementation. It is, in essence, a calendar of events and activities. The first column indicates the main activities, and for each activity, there is a suggested week/weeks in which to be implemented. When preparing a project timeline, teachers should consult their school calendar and make sure that important or major project activities don't conflict with school events or extracurricular activities. The timeline should be flexible enough to allow teachers to integrate the project activities into their units. The timeline should allow for extra time in case there are unexpected delays or challenges in implementing project activities. The project timeline should be reviewed and discussed prior to implementing project activities, and important deadlines should be identified and recorded for reminders. The estimated duration of any project will vary from classroom to classroom. It is recommended that teachers in PBSI devote between 1.0 to 1.5 hours each week to check in with students to make sure that students' activities are consistent with the timeline. Where possible, the project timeline should be communicated to school administrators and parents to keep them informed.

Chapter Summary

In this chapter, the reader is introduced to the basic principles and concepts of PBSI and the project cycle (the various stages that teachers and students must go through in order to enact PBSI). The project cycle consists of the following stages: orientation process; identifying and defining a project; planning, implementing, documenting, and reporting project findings; evaluating project learning; taking action; and proposing new projects. Teachers should not assume that all students will be interested and/or actively participate in PBSI. They need to conduct an orientation process, where they find out what their students' conceptions and misconceptions about projects, and make sure that they are adequately addressed. They need to emphasize that in PBSI environments, students take responsibility for their own learning, while their teachers act as mentors, facilitators, and resource persons. Although there may be different ways of identifying and defining a project, the chapter provided guiding

Table 3.3. An Example of a Project Timeline.

Project Activities	September					October			
	Wk1	Wk2	Wk3	Wk4	Wk5	Wk6	Wk7	Wk8	Wk9
1. Project planning: orientation to PBSI, identifying a question, developing a project plan, instrument development, testing and sampling	✓	✓	✓						
2. Pre assessment		✓							
3. Formative assessment		✓	✓	✓	✓	✓	✓		
4. Data collection	✓	✓	✓	✓	✓				
5. Journal and record keeping	✓	✓	✓	✓	✓				
6. Data analysis and interpretation				✓	✓	✓	✓		
7. Journal and record keeping				✓	✓	✓	✓		
8. Report writing					✓	✓	✓	✓	✓
9. Journal and record keeping					✓	✓	✓	✓	✓
10. Group presentations							✓	✓	
11. Post assessment								✓	
12. Assessment of project process and products/artifacts, collective reflections (summative assessment)								✓	✓
13. Taking action							✓	✓	✓
14. Future projects							✓	✓	✓

questions that could help facilitate the process. The chapter also touches on the alignment between the NSES, NGSS, and PBSI. The chapter provides a template for developing a project plan and an example of a project timeline, all of which are the cornerstone of PBSI.

Food for Thought

The following are questions that will help the reader reflect on the chapter.

1. What is the project cycle?
2. Identify and list a few roles and responsibilities of teachers and students in a PBSI environment.
3. Explain the main steps or process that students go through from identifying a project to developing a timeline.
4. How are the NSES, NGSS, and PBSI connected?

· 4 ·

IMPLEMENTING PROJECT-BASED SCIENCE INSTRUCTION

Chapter Overview

This chapter will describe the six strategies of implementing project-based science instruction: (1) teacher-centered, (2) student-centered, (3) one teacher-one classroom, (4) multiple teachers-multiple classrooms, (5) teacher-student-scientist partnership, and (6) extracurricular activity (i.e., PBSI clubs/societies and afterschool programs). It will discuss microcomputer-based laboratories tools as resources for supporting students' project work. The chapter will also cover techniques and procedures for collecting, analyzing, interpreting, and reporting different types of data that students typically encounter in project-based science classrooms.

What Factors Should We Consider Before Implementing PBSI?

Before the discussion on implementation strategies, it is important to first address the contextual factors that influence PBSI. Contextual factors that influence PBSI are many and varied. However, the following factors should be considered before implementing PBSI: type of principal; teacher prior knowledge, skill, and disposition; type of curriculum; availability of funding; level of parental involvement; level of technology adaptation; the physical environment; and time. Any one of these factors can either enhance or impede the implementation of PBSI.

Type of Principal

Although school principals perform many different functions, their main role is to provide administrative and instructional leadership to their school's communities. It is the responsibility of the principal to coordinate the functioning of all the components of the school. If a principal is inexperienced, incompetent, and/or unsupportive, the school will not function smoothly and serious problems in teaching and learning will arise. In addition, goals and outcomes may not be accomplished. On the other hand, if a principal is experienced, competent, and supportive of faculty and staff; open to innovative ideas; and willing to take a risk, it is highly likely that she or he will allow teachers to experiment with PBSI or provide support for those teachers who want to try it out. Experience has shown that in schools where project-based science instruction was implemented, usually there is a competent and supportive principal working to make this happen (Raizen & Britton, 1997). It is important to add that project-based science instruction will not work in schools where the political climate is not favorable. For instance, in a school where there is in-fighting among faculty or between faculty union and the administration over allocation of resources or over school policy issues, implementing PBSI will be a major challenge or almost impossible.

Teacher Knowledge, Skill, and Disposition

According to Park Rogers and colleagues (2011), "Implementing PBL is an extremely challenging endeavor for any teacher as it requires making changes in all major aspects of teaching—the curriculum, the instructional strategies, and the roles of both the teacher and students in the learning process" (p. 906). To implement project-based science instruction in a school, teachers should be trained in this method, in addition to their regular training in pedagogy and content area. They should know how to manage students from diverse backgrounds and learning styles, know how to mentor students as they work on their projects, and know how to assess students' projects. They should be resourceful, innovative, and willing to experiment with new ideas. Project-based instruction and learning entail some amount of responsibility on the part of both teachers and students. The responsibilities of teachers in a project-based classroom are captured in the following comments:

> The teacher played a central role in facilitating collaboration in this study. In this middle school classroom, the teacher's sensitivity to the amount of direction students needed appeared critical in helping students acquire interdependency in their groups and taking responsibility for their own learning. Although too much teacher direction deterred collaboration, too little guidance also stalled the students' progress in designing and carrying out investigations. ... It was also important to let groups make decisions on their own in order to avoid undermining the group's reliance on their team members. ... Knowing when to give support to students and when to leave students on their own, demands continuous assessment of group productivity. (Crawford, Krajcik, & Marx, 1999, p. 720)

The above quote indicates that under the project-based method, teachers may have to reinvent themselves. They will be expected to take on a more proactive and dynamic role; be teachers as well as learners; and be responsible for organizing the learning environment, providing advice on project selection, planning, and implementation. They will also mentor students throughout the process and help them assess the outcomes of their projects. This may seem to be a lot for some teachers, but it is not. This is because the students also share in the responsibility. For instance, before students start their projects, they should develop their own plans. In addition, they should discuss with their teacher how their work will be assessed to determine their learning, as well as present their final product to their peers. This shared responsibility in teaching and learning makes project-based instruction different from traditional methods of science instruction.

Type of Curriculum

The term *curriculum* here is used to mean all the things that occur in the course of instructional planning, teaching, and learning in a school (Tyler, 1981). The type of curriculum in a school can hinder or promote project-based science instruction. For instance, in a school where the curriculum requires teachers and students to follow strict standards and use certain instructional procedures, project-based science instruction will not work or thrive. However, in a school where teachers and students are allowed to implement instructional goals based on their specific conditions, project-based science instruction is more likely to work and thrive.

The science curriculum in elementary and middle schools is more flexible than in high schools. This is because the goals of science education at elementary and middle school levels are to build positive attitudes, generate curiosity about one's own environment, develop science process skills that are essential for survival in the real world, and build a foundation for future science learning (Victor & Kellough, 1997). These goals tend to lend themselves well to project-based science teaching and learning. At the high school level, the goals of science education are quite different and tend to be driven by the need to prepare students for college entrance examinations. However, it is important to note that there are some high schools where the curriculum is flexible enough to allow teachers to teach using projects. In fact, within the past two decades, PBSI curricula have been piloted and/or implemented in different K–16 science classrooms with varying degrees of success and challenges (e.g., studies by Cook & Weaver, 2015; Kiefer, Bucholtz, Goode, Hugdahl, & Trogden, 2012; Brickman et al., 2012; Duncan & Tseng, 2011; Alozie, Moje, & Krajcik, 2010; Rapp, 2008; Schnieder, Krajcik, & Blumenfeld, 2005; Marx et al., 2004; Reiser, Krajcik, Moje, & Marx, 2003). One interesting study worth mentioning here is by Schneider, Krajcik, Marx, and Soloway (2002), which found,

> PBS students scored significantly higher than students nationwide on many items [on the *12th-grade 1996 National Assessment of Educational Progress* (NAEP) science test]. Even compared with groups that traditionally score higher on achievement tests (middle-class and White students), on average the PBS students, including minorities, outscored the national sample on almost half the items. Also, it is not known whether any students in the national sample participated in a PBS program; therefore, the national sample is not necessarily a non-PBS group. Still, this PBS group of 10th- and 11th- grade students performed higher on this set of 34 items than the national sample of 12th-grade White students. (p. 419)

Availability of Funding

Project-based science instruction demands that resources be made available for students to carry out the projects of their choice. This means that tools, materials, equipment, hardware, and software should be available to all students. Professional development may also be needed to provide teachers with the skills they need to implement project-based science instruction. All of these require funds. Where funds are not available, it will be very difficult to implement project-based science instruction. Sometimes it is possible to implement PBSI in a school where funds are limited, because the types of projects the students select may not require resources other than those that already exist in the school, such as a computer lab, software, reliable Internet access, digital cameras, and video cameras. Sometimes, teachers are willing to take the initiative and try it out at their own expense. Speaking from personal experience, this author has, on a few occasions, purchased materials out of pocket to support a PBSI idea in his classes.

Parental Involvement

Project-based science instruction requires students to conduct projects over extended periods, which means sometimes doing work outside of school. The implications for this are that students will need parental or guardian support to pursue projects to their end. Parental involvement does not only mean attending parent-teacher meetings or conferences, but also having high expectations for their children, supporting them at home, advocating for them, and holding them accountable. In a school where there is an active parental involvement program, project-based science instruction will thrive because students will presumably get the outside support they need to complete their projects. Parental involvement in education is considered to be an influential factor in students' overall academic outcomes (Colley, 2014; Greene, 2013; Epstein, 2010; Delgado-Gaitan, 2001; Comer, 1980).

Level of Technology Adaptation

Technology can make our lives simple or complicated. However, in schools where "appropriate" technologies exist, as well as the right training and support service, important educational goals could be achieved. Project-based science instruction is likely to work well in schools where the teachers and students have access to computers, software, servers, the Internet, audio recorders, video players-recorders, CD-ROM write-rewrites, scanners, digital cameras, digital camcorders, microcomputer-based laboratories, calculator-based laboratories, environmental monitoring tools, microscopes, laboratory apparatus, and measuring instruments. This is because these technologies provide opportunities for creativity, experimentation, and collaboration and tend to enhance the teaching and learning process (Tinker & Papert, 1989).

Physical Environment

The internal and external environment of a school can either promote or hinder the implementation of project-based science instruction. For instance, in a school where the ventilation and toilets do not work and/or the classrooms are overcrowded with dilapidated furniture, project-based science instruction may not work as well as it should. This is because teachers and students, instead of being able to concentrate on the teaching and learning process, have to deal with the elements of the internal environment, which influences their educational productivity.

The external environment of a school includes its recreational fields, open space, trees, wildlife, and access to an ocean, a river, a mountain, a national park, a major university, a public library, a museum, or a big city. Schools located in an external environment with one or more of the features listed are more conducive for project-based science instruction. This is because such external environments provide ample opportunities for students to explore and test ideas.

It is important to bear in mind that teachers and students are also capable of influencing their environments. And since this is the case in some schools, it should be noted that project-based science instruction might still work even in schools where the internal and external environments are less desirable. However, the message here is that the environment is an important factor to consider when thinking about implementing project-based science instruction in schools.

Time

Implementing project-based science instruction requires time in terms of instructional planning, scheduling activities, developing collaborative relationships, implementing activities, supervising students' team work, assessing students' projects, and learning to use technology appropriately. When time for all these activities are taken into account, it adds up and could be overwhelming for teachers and students. Proper scheduling and management of teachers' and students' time is a critical factor to be considered when implementing project-based science instruction (Polman, 2000; Mistler-Jackson & Songer, 2000). However, according to Alozie, Ecklund, Rogat, and Krajcik (2010), "The time required to effectively enact project-based materials is greater than with traditional learning methods, but students retain more knowledge and are able to apply it to new situations" (p. 229).

Strategies for Implementing Project-based Science Instruction

Teacher-centered

Project-based science instruction could be implemented using a teacher-centered strategy. This means that the teacher will be the principal investigator while the students serve as co-investigators. In this strategy, the teacher is responsible for coming up with the investigative question, developing the plan for conducting the project, and guiding and mentoring students throughout the implementation of the project. Teacher-centered implementation of PBSI has its advantages and disadvantages and should be conducted with care. One advantage is that it allows the teacher to guide students and to ensure that PBSI is aligned with curriculum requirements and science learning standards. A disadvantage is that this type of implementation strategy has the potential for the teacher to exert too much control on the learning situation, leaving the students to feel that its their teacher's project, that they have no input or stake in the project.

Student-centered

A student-centered strategy is where students take full responsibility for planning and implementing projects in their own classroom. This usually involves students formulating their own driving or investigative questions, working in small collaborative groups to develop project plans, implementing and evaluating their projects. This strategy requires that teachers take risks and allow students to connect learning to their own communities and learn at their own pace. Their roles are more of facilitators, resource persons, and mentors. According to Berenfeld (1994), "By involving students in the decision of what to study, we ensure that their studies are of intrinsic interest to them. By engaging students in examining the environmental health of their communities, we ensure that they are fueled by social concerns" (p. 205). This strategy does not work well in an environment where the emphasis is on structure, testing, and accountability.

Teacher-Student Partnership

Sometimes a teacher and his or her students collaborate to plan and implement PBSI, and when this happens, this strategy is referred to as teacher-student partnership. Similar to teacher-centered strategy, in this PBSI environment, the teacher serves as principal investigator, while the students serve as co-investigators. However, the difference is that in this situation, the teacher and students truly collaborate and share in the formulating of the driving question. This strategy is appropriate when resources are limited to allow for more than one group of students to work on a project. In addition, it is ideal for students at lower grade levels or where safety is a major issue. When the teacher is the principal investigator, it is a wonderful opportunity for him or her to demonstrate different science process skills.

One Teacher-One Classroom

In this strategy, one teacher and his or her students plan and implement a project from start to finish. The project could be student centered or teacher centered; however, this strategy does not include any kind of collaboration or cooperation with other classrooms or outside entities. The students do not work in small groups, but it is a whole classroom activity. In other words, the classroom is the group.

Multiple Teachers-Multiple Classrooms

This strategy involves two or more classroom collaborating with each other, where teachers and students all work together on planning and implementing one or multiple investigative questions. The project could be a singular project or it could be several projects, but the idea is that they are all connected. Usually the project addresses a central theme in the curriculum or a common investigative question. Multiple teachers, multiple classrooms could include classrooms in the same school, district, state, or across states or countries. It could also be one grade level or multiple grade levels. The challenge of the multiple teacher, multiple classroom strategy is that the level of classroom management is complex, while the demand and utilization of resources is high.

Teacher-Student-Scientist Partnership

This strategy is one of the most powerful strategies for implementing project-based science instruction. In this strategy, the teacher and students collaborate with a scientist or group of scientists on a long-term project. The project is usually determined based on mutual interest within the partnership.

For instance, a teacher and his or her students might be studying weather and climate in their earth science course and contact a climate scientist who is already doing research work in this area. The teacher and students usually serve as research assistants in that they help in data collection and pilot testing of tools and instruments in exchange for the scientist's mentoring, access to the project, and serving as a resource person for students. In order for teacher-student-scientist partnership (TSSP) to work, certain factors should be taken into consideration.

Evans, Abrams, Rock, and Spencer (2001) identify the following critical components of TSSP: ensuring real access to scientists; the need to have teacher professional development and workshops on methods, procedures, and protocols; opportunities for student conferences to present their projects for peer review; availability of quality support materials and research protocols; alignment of activities and projects to science content; learning standards and assessment goals; cost of participating in terms of time; funding opportunities; and assessing trade-offs. They note,

> One of the challenges faced by [TSSP] is generating the critical data required for the overarching scientific research goals, while still trying to encourage students to develop their own questions within the context of their field sites and smaller data sets. Asking questions, developing valid experimental approaches, and analyzing data are some of the most exciting of the scientific enterprise and [TSSP] should help teachers engage students in a full range of science activities. (p. 323)

TSSP, the researchers warn, must avoid a partnership that is one-sided and based on students as "data collectors." This kind of partnership is usually unsustainable and has the potential to leave the impression on students' minds that science is only about data collection.

Extracurricular Activity (PBSI Clubs/Societies and Afterschool Programs)

Project-based science instruction can also be implemented outside of regular classroom time. A teacher or group of teachers could form science clubs or societies and use this as a medium to plan and implement projects. Alternately, a teacher could also partner with a local university or not-for-profit science organization to set up an afterschool science program in which students plan and implement projects, either as a group or as individual students.

Internships

Projects could also be implemented through internships. In the internship strategy, individual students or groups of students are placed with a scientist, and the students work in the scientist's lab and learn basic techniques and procedures relating to a particular discipline. Internships could be paid or unpaid activities. Students could also get credits for it that they can apply toward fulfillment of requirements or graduation.

Using Microcomputer-based Laboratories Tools

Microcomputer-based laboratories (MBL) refer to the application of computers in science laboratory work. In the mid to late 1980s, science educators, scientists, and researchers at TERC developed and piloted different mathematics and science curricula supported by innovative computer software and hardware. During this period, they discovered that computers could serve as powerful tools for science teaching and learning, especially when they are equipped with appropriate software and probeware (Mokros & Tinker, 1987).

Probeware or sensors are simple tools that are connected to a computer or calculator during science laboratory classes to collect real-time data. They work by converting physical or chemical variables in the environment into electrical signals, which are then displayed as graphs or data tables on a computer screen. There are different types of sensors or probes, but the most common ones are temperature, motion, sound, pressure, light, and pH. Metcalf and Tinker (2004) studied the feasibility and cost of using probeware and related curriculum materials in elementary and middle school science classrooms. Using a sample of 30 teachers in the first year and 8 teachers in the second year, they piloted two curriculums that incorporated the use of six low-cost probewares. Their findings showed that probeware and handheld devices enhanced students' science investigations as it relates to "physical correlation between phenomenon and modeling" (p. 43) and addressing students' alternative conceptions in science. From their research, Metcalf and Tinker concluded,

> Our research was also concerned with the feasibility of using these materials in middle school science education. The data showed that, with minimal training either face-to-face or on-line, teachers were able to implement the units quite well. Classroom observations and post-interviews showed that teachers and students managed to succeed in almost every investigation they undertook. (p. 49)

Before MBL was introduced into science education, some science laboratory work, particularly in physics and chemistry, was not always accessible to students due to complexity, cost, and safety issues. MBL changed the way science laboratories are conducted now because, according to Krajcik and Layman (1992), "The capabilities of MBL to immediately transform data from each experiment into a graph—the most powerful information presentation, is something that has not been possible in the past" (p. 102).

Since its introduction in science education, various studies have been conducted to determine the effects of using MBL on student understanding of basic science concepts and acquisition of science process skills (Tinker, 1986, 1996; Adams & Shrum, 1990; Thornton & Sokoloff, 1990; Redish, Saul, & Steinberg, 2015). According to Tinker (1996),

> Microcomputer-based labs, the use of real-time data capture and display in teaching represent one of the most valuable innovations microcomputers have contributed to science teaching. This technology gives the learner new possibilities to explore and understand the world, and to see it represented symbolically in ways that greatly increase comprehension. (p. v)

Thornton and Sokoloff studied the effects of an MBL curriculum on kinematics—which includes motion detector software and hardware—on secondary school and college students' understanding of physics concepts such as motion and velocity. The curriculum was based on a "guided-discovery approach," and provided the students with the opportunity for collaborative group learning, while the motion detector allowed students to conduct independent investigations of their body motions, calculate position, velocity, and acceleration, and generate real-time graphs. In reporting their findings, the researchers note,

> Microcomputer-based laboratory and curriculum have the potential to help students develop a solid conceptual basis for understanding the world around them. Through the use of these materials, students' interactions with the physical world can be connected to the underlying principles that constitute scientific knowledge, thereby helping them to develop a conceptual, qualitative understanding that can be applied both inside and outside of the classroom. (p. 866)

MBL has several advantages over traditional science laboratory. For instance, it helps students develop understanding and use of data; makes it easier to understand abstract science concepts such as temperature, motion, velocity, pH, and pressure, to name a few; and allows students to experience real-time data and receive immediate feedback, manipulate variables in ways that are not possible in a traditional science laboratory, acquire science process skills, and become familiar and/or competent in the use of computer software and hardware. Like any new innovation, MBL has its own limitations. Although the prices of computers, probeware, calculators, tablets, and software are coming down, not all students may have access. MBL requires a school culture that nurtures innovation and risk taking on the part of administrators and teachers. In addition, for effective learning to occur, an MBL requires a curriculum supported by appropriate tools.

In implementing their projects, students are expected to and should use tools, technologies, and materials; collaborate in carrying essential tasks such as planning, data collection, and preparing of reports; calibrate and use tools to collect, analyze, and interpret data; and learn basic concepts that underpin the questions and problems they are studying. MBL is pedagogically consistent with PBSI because its fundamental goal is to put students front and center in the science classroom. It empowers students to take charge of their own learning, whether as individuals or as groups. It is a cost-effective way to implement PBSI because it is multipurpose and can be used to study more than one question, problem, or discipline. There are various MBL curricula and tools available, some on the Internet as free, open-source software, and others for purchase from vendors. Some of the most popular MBL products on the market come from two major vendors-companies, Pasco Scientific (www.pasco.com) and Vernier Software and Technology Company (www.vernier.com). These companies have a long history of working with schools and colleges and developing tools and curricula to support MBL. Visiting their websites is a good place to start for those teachers who are interested in incorporating MBL in their PBSI.

Collecting, Analyzing, and Interpreting Project-based Data

This section of the chapter will review and discuss some of the issues, methods, procedures, and challenges pertaining to collecting, analyzing, and interpreting students' project data. The following questions will be used as a guiding framework: (1) Within the context of PBSI, what do we mean when we say "project data"? (2) How do teachers and students collect project data in PBSI environments? (3) What are the most common types of project data that teachers and students encounter in PBSI? and (4) What are the main methods or procedures teachers and students use to analyze project data?

The term *project data* is used to mean any quantitative or qualitative information that students collect to answer a driving project question. Project data could be in the form of numbers, words, text, or images and could be in digital or nondigital format. The methods that students use to collect project data vary depending on their driving research questions, content area of study, theme, topic, grade levels, availability of resources, and other contextual factors. However, in general, students use the following instruments or tools to collect project data: observation forms or protocol, standardized data sheets, microcomputer-based laboratory tools, such as probeware or sensors, surveys or questionnaires, digital cameras or videos, scientific instruments (e.g., rulers, measuring tape, pH meters, thermometers, Global Positioning Satellite (GPS) tracking devices, graphing calculators, telescopes, etc.), and portable weather stations, to name a few. Tables 4.1–4.5 provides examples of qualitative and quantitative data collection instruments that were used in a PBSI focusing on earth science/environmental sciences for middle and high school students.

Table 4.1. Soil Characteristics and Types of Tests You Can Do.

Soil Characteristics	Type of Test You Can Do
Soil Types: Is it clayey, sandy, or loamy (i.e., mixture of clay, sand, and organic matter)?	**Visual:** Look at the soil. Sand is usually light brown, white, rusty, or mixed. Clay is usually gray or ash. loam is usually dark or black. **"Wet and Feel Test":** Clayey = slippery, Sandy = gritty or coarse, Loamy = medium or smooth
Color	Visual
Texture: Is it fine, coarse, or medium?	Visual and "Feel Test"
Structure: Is it loose, crumbly, granular, or blocky?	Visual and "Feel Test" Take a hand full of the dry soil sample and try to roll it into a ball. Observe and describe your result.
Organic matter content: Is it low, medium or high?	Visual Is it dark or light in color? High organic is associated with dark or black soils.
Drainage capacity: How slow or fast does it absorb water?	"Wet Test" Measure water, pour water on soil, and record time it takes to absorb (milliliters per second).
Density: How compact or hard is the soil?	"Pencil Test" Take a pencil and push it into the soil. Does it enter easily? (1) Does it require force? (2) Does it require a hammer or tool to push it in? (3) 1 = no compaction, 2 = low compaction, 3 = moderate to high compaction

Table 4.2. Data Sheet for Soil Testing.

Sample:_____

Date:_____

Soil Characteristics	Descriptions
Soil Type:	
Color	
Texture:	
Structure:	
Organic matter content:	
Drainage capacity:	
Density:	

Table 4.3. Data Sheet for Soil Testing–1A.

Site/Location:_____

Date:_____

No.	Description of Soil Sample	pH	N	P	K		
Maximum							
Minimum							
Mean							
Standard deviation							

Table 4.4. Water Characteristics and How to Describe Them.

Water Characteristics	How to Describe
Color	Color of the water may be caused by impurities. Iron can cause color to be brown, algae, greenish; organic matter, dark brown.
Odor	Describe as fishy, rotten egg, swampy, or fresh.
pH	A measure of the hydrogen ion concentration. Tells you how acidic or basic the water is. pH 1–4 very acidic, pH 7 neutral, pH 12–14 very basic. Most water plants have a pH of 6.5–12, warm water fish pH of 6.0–9, cold water fish pH of 6.5–8, most insects, pH 6–8.5.
Dissolved Oxygen	Dissolved oxygen is a measure of the health of living things in water. Oxygen is required by both aquatic and terrestrial animals. Different factors affect the supply of oxygen in water. These include temperature, light, dead organisms, stream flow, and human activities. Less oxygen in the water means fewer organisms in the water.
Nitrate	A measure of the concentration of the nitrate ions. Main sources are agriculture, urban runoff, livestock industry, sewage treatment plants, industrial waste, and automobile emission. Levels in fresh water are between 0.1 to 4 mg/L.
Turbidity	How clear is the water? Is it cloudy? Are the particles in suspension?
Temperature	Measure of hot or cold, affects solubility of oxygen, photosynthesis, and survival of aquatic organisms; acceptable levels varies.

Table 4.5. Data Sheet for Water Testing.

Water Source:_____

Date:_____

Water Characteristics	Date	Highest Value	Range	Unit
Color				
Odor				
pH				
Dissolved Oxygen				
Nitrate				
Turbidity				
Temperature				

Based on a review and synthesis of the research on PBSI in chapter 2, the most common types of project data that teachers and students encounter in PBSI environments are one sample, one variable and multiple samples, multiple variables types of data. Most of these data are continuous data. Continuous data follow the number line without regard to category. For example, data in the form of –3, –2, –1, 0, 1, 2, 5, 4, 8, 9, 13, 20, 7 … infinity; percentages; test scores; daily precipitation; or rainfall during a specific month. An example of one sample, one variable data is the hourly temperature in a 24-hour period or sample of pine tree seedlings and their heights. Multiple samples, multiple variables types of data are project data that consist of two or more variables with several levels of measurements. For instance, colony collapse disorder (CCD), the rapid loss of adult worker bees in bee hives in many parts of the United States in recent times, is a challenging and worrying problem for beekeepers and scientists alike. Although scientists have not agreed to the real causes yet, the theory and/or the current research available point to parasites, pathogens, pesticides, destruction of habitat, and other environmental stressors as causes. A potential study to investigate this problem will include collecting multiple samples, multiple variables types of data on the following: apiaries, brood condition, parasites, pathogens, habitat, and environmental stressors by locations and/or seasons for both control and treatment groups.

Beside one sample, one variable and multiple samples, multiple variables types of data, students in PBSI sometimes collect categorical data, which is data that can be placed in fixed order, such as race, gender, days of the week, types of oak trees, forest canopy, or the electromagnetic spectrum. Categorical data could be either ordinal (dichotomous as in yes or no) or nominal (more than two discrete categories). A common example of categorical data is data from surveys that consists of rating scales (e. g., 1 to 5, where 1 means strongly disagree and 5 means strongly agree). Other forms of project data could be qualitative in nature, such as interviews, field notes, observation notes, photos, videos, and/or drawings. Figure 4.1 presents a decision tree for analyzing project data. A decision tree is a first aid kit for making decisions about complex situations, issues, or questions. As can be seen from the figure, there is an overarching question about data. Depending on the response to the question, the students and teachers are directed to the next option, and then to the next option, until a final decision is reached on what methods and techniques of analysis to use and when to use them. The decision tree has its own limitations. For instance, it does not provide detail explanation on how to conduct the data analysis or how to interpret the results of the analysis. One of the responsibilities of teachers in PBSI environments is to help students make sense of data. This means that teachers themselves should be well prepared and experienced in procedures for collecting, analyzing, interpreting, and reporting different types of scientific data. Teachers should have an understanding of basic research design and statistical methods. This could come from teacher preparation course work or professional development during in-service teaching.

It is important to note that there are different methods and procedures for analyzing project data, and it is beyond the scope of this book to discuss them in detail. What is presented in figure 4.1 is a simplification of a very complex system. It is hoped that this decision tree will serve as a beginning step, and to gain further information on data analysis, the reader is referred to chapter 8, which includes various resources that could facilitate knowledge and understanding of data collection, analysis, and interpretation in PBSI environments.

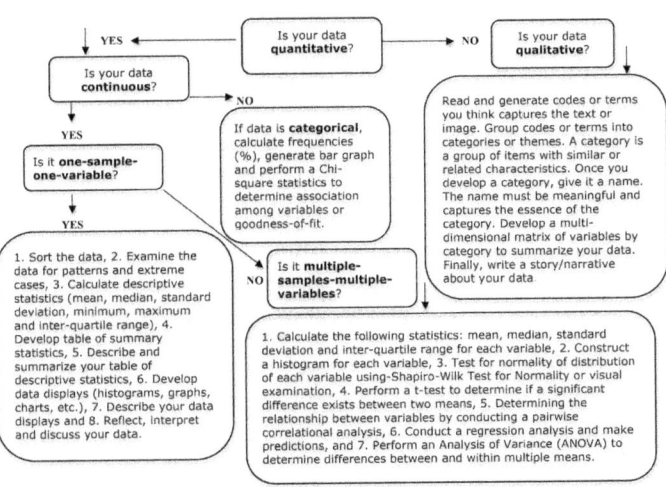

Figure 4.1. Decision Tree for Analyzing Project Data.

Preparing and Presenting Project Reports

It is generally understood that when students conduct projects, they should have reports or papers at the end of their activities that they present for peer review. Project reports or papers take different formats and may include different content. However, the expectation in PBSI is that project reports or papers document in full detail all aspects of the project, including and not limited to the following: the background and context, significance of the project, driving question, prior research on the driving question, method of study, findings, limitations, and follow-up action to be taken (implications for policy and practice).

There are different approaches that teachers in PBSI use to help their students develop report writing and presentation skills. One approach is to have students study the "anatomy of a scientific journal articles" (Schinske, Clayman, Busch, & Tanner, 2008, p. 56). This approach involves students identifying, selecting, and conducting a dissection of samples of articles by looking at their abstracts, research questions, methods, figures, tables, results, discussion, and references. The idea is that by having students examine the different parts of peer-review articles, their purposes, and how to deconstruct and construct them, they will learn how to prepare their own articles.

Another approach is to have students prepare posters of their project. Poster presentation requires limited resources such as poster paper and markers. Students can work on their poster presentation in groups or as individuals. Posters are a good way to summarize the processes and outcome of projects. Students can also do them at home and bring them to school for presentation. Recently, the poster presentation has been transformed into a new form, the digital presentation, using Microsoft PowerPoint (MPP). MPP is a software that allows users to create a multimedia presentation that incorporates text, voice, sound, still images, and video. PowerPoint presentations are easy to learn and produce. When used appropriately, MPP could provide an opportunity for students to learn technology skills and, at the same time, present their project work without the challenges of writing a full project report.

As already noted, in chapter 5, Vacchina and Aguirre discuss alternative ways in which their students conducted their project presentations. According to the authors, in addition to writing articles for their local newspaper, their students identified and selected photos, video clips of project activities and other artifacts, and created DVDs and a short movie on their project. The movie was screened on local television station for public viewing. Some computers are outfitted with speech recognition software, and where this is available, it could be a useful tool for those students who are not good in word processing, especially as they prepare their project reports.

Finally, a good resource on format and style for preparing project reports or research papers is the American Psychological Association (APA) *Publication Manual* (APA, 2009). Teachers could also refer to *The Craft of Research*, by Booth, Colomb, and Williams (1995), published by the University of Chicago Press. It is important to note that there are several resources on the Internet that provide information on format and style, and the reader is encourage to Google the topic to learn more. Below is a simple template for preparing a project report or paper, which could be modified to fit different contexts.

Title of Paper, Name of Author(s), Affiliation, and Date

INTRODUCTION OR PURPOSE OF STUDY
In 1–2 pages, provide background or context and explain the purpose and significance of your project (study). State your research question, define/explain your variables. State your hypothesis (if any).

PROCEDURE OR METHODS
In 1–2 pages, explain or list the main steps you have taken to conduct your project, including sampling, data collection instrument and procedure, and data analysis procedure.

RESULTS AND DISCUSSION
In 2–3 pages, summarize and describe your results (data). What do they tell you? What are they not telling you? What are the plausible explanations for your results?

CONCLUSIONS
In 2–3 pages, respond to the following questions: What conclusions could you draw from your results? Were you able to answer your research question with your data? Are your results conclusive? What are the limitations of your results? How would you address the limitations in the future? What is your next step?

REFERENCES (If applicable)
Use APA style.

APPENDICES (If Applicable)
Title each appendix appropriately. Place in order they are cited in report.

Figure 4.2. Template for Preparing a Project Report.

Chapter Summary

In this chapter, the six strategies of implementing project-based science instruction—teacher-centered; student-centered; one teacher, one classroom; multiple teachers, multiple classrooms; teacher-student-scientist partnership;

and extracurricular activity (i.e., PBSI clubs/societies and afterschool programs)—were described. The chapter identified and discussed the factors to consider before implementing PBSI. In addition, the role of MBLs as resources for supporting students' project work was discussed. The chapter presented various data collection and analysis templates that could be used to support students as they collect, analyze, and interpret project data. A decision tree on the different types of data and the techniques and procedures for analyzing them are provided, as well as an overview on preparing and presenting project reports.

Food for Thought

The following are questions that will help the reader reflect on the chapter.

1. Identify and describe the six strategies of implementing PBSI.
2. What factors should teachers consider before implementing PBSI?
3. What are microcomputer-based laboratories? Provide one example of how it could be applied in the PBSI classroom.
4. Create a decision tree for analyzing a one sample, one variable data set.

· 5 ·

CASE HISTORIES OF PROJECT-BASED SCIENCE INSTRUCTION

Chapter Overview

In this chapter, four cases or examples of project-based science instruction will be presented and discussed. The four cases will focus on the following subject areas: biology, chemistry, earth and environmental sciences, and physics. Each case will be discussed with a particular emphasis on background and context, teacher's role, students' roles, implementation strategies, project activities, assessment of learning, challenges, and lesson learned.

Case 1. Project-based Biology: The Herpetology Project

Background and Context

Herpetology is defined as the scientific study of amphibians and reptiles. The Herpetology Project was started in the mid-1970s by the U.S. Geological Survey and the Florida Caribbean Science Center to "assess the status and distribution of amphibian and reptiles in the threatened Florida biotic communities" (USGS, 2000, p. 1). Since its creation, the project has been expanded to include a wide range of basic and applied research activities on amphibians and reptiles in the Southeastern United States and the Caribbean. The Herpetology Project, in collaboration with other nongovernmental scientists, have developed and implemented a variety of tools, techniques, and research protocols for the study of amphibians and reptiles. These tools, techniques, and research protocols include the "implementation of biometrically-based community inventory and monitoring programs, population modeling, telemetry (radio and satellite), genetic analysis and database management" (p. 1).

In 2004, students at Hudson High School in Hudson, Massachusetts, and their science teachers decided to start their own herpetology project after participating in a science field trip to the Pantanal in Brazil, hosted by the Earthwatch Institute. Two events in their school converged that led to the students and their teachers to make this decision. One year after the students moved to a newly constructed building and softball field, they noticed that a turtle from a river nearby, the Assabet River, had traveled all the way to their softball field and laid its eggs. "Exactly one year later, the event occurred again in almost the same location" (Vacchina & Aguirre, 2008, p. 51). Because the students already had some experience studying amphibians and reptiles as part of their field trip to Brazil, and because of the need to protect the turtle and its eggs, the students and their teachers decided to focus their environmental, science, environmental chemistry, and biology classes on the Herpetology Project. The project provided students with the opportunity to catch, track, study, and protect amphibians and reptiles in the Assabet River ecosystem. According to Vacchina and Aguirre, the purpose of the project was to investigate "how the increased use of playing fields might affect the nesting patterns of the turtles (snapping, painted and common musk)" (p. 51).

Teacher's Role

From the available documents on this case, the role of the teachers is not fully discussed. However, this could be inferred. The roles and responsibilities of the teachers in this case included curriculum and unit planning, classroom management, assessment, grant writing, development of students' research, and technology skills. Prior to implementing such a project, the teachers reviewed their school, district, and state curriculum guides to make sure that their project goals and objectives were aligned. The teachers were also responsible for providing instruction and guidance of content knowledge as it related to the curriculum guides and national science education standards. In addition to curriculum-unit planning, the teachers were responsible for creating a learning environment that promoted active participation of all students regardless of background or status. The teachers supervised research activities, such as design and construction of instruments, data collection, data analysis, preparation of reports, and presentation of reports. They were responsible for time and resources management. Assessing students' learning in a project-based science setting could be challenging because of the complexity of teaching and learning going on. This requires the use of performance-based assessment methods and techniques. In this case, the teachers assessed students' learning through

> a field-research journal of reflections on their research and its impact on local environmental issues, as well as scientific thinking in more global terms. In addition, students write articles for a summer edition of the district newsletter … document an extensive amount of information during the project. They take designated digital photographs, videotape activities, collect positional data using a GPS device, and use mapping software to plot data on a map of the area. … A DVD of photos and video footage are woven into a live television production. (Vacchina & Aguirre, p. 54)

As already indicated, a project such as the Herpetology Project requires resources, and when they are not funded by the school, the funding usually comes from outside sources. This means that the teachers are also busy with writing grants and organizing fundraising activities as confirmed in the following words: "We have been fortunate to receive grants to help offset our costs—five from our community education foundation (two of which were results of student-written proposals) and one generated by students through the sale of candy at lunch" (Vacchina & Aguirre, p. 55). In order for the students to collect, analyze, and report their research project results, they needed to be instructed, provided with hardware and software, and given the opportunity to practice their technology and research skills by their teachers. Throughout this project, it was the responsibility of the teachers to monitor and enforce safety protocols with regard to the handling of live animals, proper uses of tools and equipment, wearing of appropriate and protective clothing, and securing of permits from the relevant state or local authorities to capture and release protected live animals.

Students' Role

The students played a variety of roles and took responsibility for carrying out the main research tasks involved in the project. For instance, the students served as coprincipal investigators, engineers, biologists, chemists, environmentalists, analysts, videographers, photographers, geographers, writers, reporters, TV producers, advocates, and multimedia experts. In addition, they implemented the project from beginning to the end. Some of the main research activities the students engaged in were as follows:

Instrumentation. The students constructed the pitfall traps to collect the turtles. The traps consisted of four holes deep enough to fit a 30-liter bucket, spaced 10 meters apart and 40 meters long. A fence along the 40-meters-long boundary, consisting of black plastic sheeting 1 meter high, was then constructed. A total of four of these traps were constructed at different locations between the river and the school. Students used a GPS device to mark the location of each bucket.

Data collection. Students checked the traps daily, removed turtles that fell into the traps, engraved them for identification, implanted microchip under teacher supervision, and provided fresh water to prevent dehydration. In addition, students took photographs of different views of the turtles, measured them, weighed them, and released them.

Data analysis. Students conducted descriptive statistical analysis to look at trends and formulated a hypothesis to explain the increase in the number of turtles despite the fact that their ecosystem had been disrupted.

Reporting writing. Students kept field notebooks or journals in which they documented their daily activities, observations, and reflections throughout the implementation process. In addition, they collaborated in authoring articles for the summer issue of their local newspaper. At the culmination of the Project, the "students peer edited PowerPoint presentations and made DVDs of graphs, spreadsheets, photograph collections, digitized animal sounds, and video clips taken during their research to communicate their findings to the school and community" (Vacchina & Aguirre, p. 54). The project was also reported in the local media. For instance, the *Boston Globe* (2007) ran a story titled "The Riverbank Is Their classroom: Hudson Assignment Benefits Turtles." In this story, the paper noted,

> That curriculum includes daily trips to the traps to retrieve the animals. Those collected are brought to the classrooms, where they are measured, weighed, photographed and in some cases, branded. The traps have resulted in the capture of 10 bullfrogs, 40 green frogs, 6 pickerel frogs, 5 American toads, 10 painted turtles, 6 snapping turtles, 5 musk turtles and a spotted turtle. (p. 2)

Presentation of project findings. Students produced a live TV segment that was "broadcast over the local cable station and posted online to educate the community about the local wetlands" (Vacchina & Aguirre, p. 54).

Strategies for Implementation

The strategy employed in this case could be described as student-centered because the students played a major role in planning and carrying out the project. They took full responsibility for the project, while their teachers played mentoring, facilitating, and supporting roles. The teachers also served as resources as well as providing technical assistance to their students on matters relating to science content and alignment with standards, identification, and use of scientific tools and technologies for their projects.

Project Activities

This project lasted multiple years (more than four) and involved collaboration among interdisciplinary subjects such as biology, environmental science, and environmental chemistry. Although not explicit in the documentation on this case, from the information available, one can safely assume that this was a long-term project that lasted several school terms with different groups or classes succeeding/inheriting the project from the other. Activities varied from daily to weekly, biweekly, or monthly. Below is a summary-list of activities that students were engaged.

1. Project planning
2. Establishing protocols
3. Securing permission for trapping turtles
4. Constructing turtle traps
5. Capturing, engraving, measuring, weighing, photographing, videotaping, implanting microchips, and tracking amphibians and reptiles
6. Learning computer software and hardware
7. Analyzing data, writing articles, creating digital video reflections, producing a live TV show, presenting on local TV, making DVDs of graphs, photos, and video clips
8. Plotting using GPS software and mapmaking
9. Assessment of learning
10. Community service and advocacy

Challenges and Lesson Learned

The teachers reported that a major challenge and lessons learned from this project was time management. As they put it, "Taking on a research project such as this requires students and teachers to commit time and energy above and beyond what is typically required of a regular school day … keen time management skills are important to the success of such an undertaking" (Vacchina & Aguirre, p. 55). Another challenge was cost of materials and equipment, which were purchased through grant monies. Having a dedicated storage space to store tools, equipment, and materials during the implementation of the project was also cited as one of the challenges.

Case 2. Project-based Chemistry: Investigating Carbon Dioxide and Indoor Air Quality

Background and Context

In 1992–1994, I worked at TERC in Cambridge, Massachusetts as a curriculum writer and director of evaluation of the Global Lab Project (GLP). During this period, I witnessed project-based science instruction live and saw its impact on science teachers and their students who participated in the project. The GLP was launched in 1990 by TERC, with funding from the National Science Foundation. The purpose of the project was to transform the way science was being conducted in U.S. science classrooms. Participants in the GLP included teachers and students from high and middle schools across the United States and abroad, as well as practicing scientists. Each participating school received scientific tools for environmental research, access to a telecommunication network, and state-of-the-art written instructional materials.

The underlying thinking behind the GLP curriculum was that giving students access to a telecommunications network and scientific tools would enable them to collect data on local and global environmental problems that they found interesting and compelling; that collecting and sharing of data online via a computer network would build a community of students, teachers, and scientists with similar interests; and that a community of people with similar interests, guided by experienced teachers and scientists, would be able to create the social context essential for the authentic pursuit of scientific knowledge. Becoming part of this community would allow students to experience science learning in a way that was meaningful, relevant, interdisciplinary, technology supported, and collaborative.

The implementation plan for the GLP was divided into two phases. I was involved in the implementation of both phases, as a curriculum developer and as director of evaluation. The first phase (GLP I: 1990 to 1992) was a period marked by the development of instructional materials, scientific tools, and field testing with a wide variety of unstructured student-based research projects. For example, during this period, project staff, including myself, in collaboration with scientists, teachers, and scientific companies, developed and field-tested a "Toolkit" to support the GLP curriculum. The Toolkit included a wide range of low-cost environmental monitoring tools such as the TERC Rubber-Thread Ozonemeter; the TERC Solar Altitude Gnomon; an airborne particulate collector; TERC field interface; temperature and light probes; TERC Air Pump-gas sampling tubes for carbon dioxide, carbon monoxide, and nitrous oxides; the TERC Soil and Water Test Kit; an inclinometer; a lichens bioindicator chart; a cloud chart; a particulate chart; a small 100-power microscope; an audio tape recorder; and a camera. A curriculum guide called the *GLP Notebook* was also developed for students and teachers who participated in the project. The *Notebook* focused on local and global environmental monitoring activities. In addition, it included a methodology for teaching and learning science based on the constructivist paradigm. A community of teachers, students, and scientists was formed during this time and a computer network for sharing experience established using EcoNet.

The second phase (GLP II: 1992 to 1994) saw the implementation of a structured curriculum backed by the Toolkit, the network, and computer software for networking. The GLP II curriculum taught science across two semesters. In the first semester, participants carried out the following instructional procedures: Site Selection, a process whereby students and their teachers identified and selected a study site in which they carried out their research projects. After selecting a study site, students conducted an Eco-Inventory of their study site. In the Eco-inventory phase, students evaluated the environmental conditions of their study site by conducting qualitative tests of the soil, water, and air quality. They observed the fauna and flora to determine the environmental health of their site. In addition, students also asked questions and formulated preliminary hypotheses. They documented their findings and shared them with other schools through the network.

Following this phase was Eco-Profile, which required students to make quantitative measurements on selected parameters of their study site. They also participated in a series of procedures called Environmental Snapshots, a number of scientific activities in which all GLP schools took part from September to December 1993. All activities were conducted on the same day using the same tools and techniques over five continents. This phase introduced the students to various scientific tools, the concept of calibration, and standards of measurement. In addition, students also learned data sharing and analysis, facilitated by the TERC telecommunication, data analysis, and the mapping software Alice.

In the second semester, participants carried out the final phase, called Eco-Research. This was the phase in which students integrated all the science content and process skills they had learned during the first semester. Specifically, students selected one of the following research strands: Biodiversity and Field Exploration, UV and Stratospheric Ozone, Weather and Climate, Your Classroom Environment, The Air We Breathe, Environmental Chemistry, Ionizing Radiation, The Mysteries of Animal Migration, and What's in Your Water. They then chose a research problem and designed and conducted a study. During the process of their study, students learned about the nuts and bolts of working together on a scientific project through actual collaboration with other students and scientists in data collection and analysis. Finally, students prepared and presented their report in *The Planet*—a journal where GLP participants published their work—and took action.

Carbon Dioxide and Indoor Air Quality Project

In 1993, a teacher and her students in a GLP middle school in Texas decided to study carbon dioxide and classroom air quality in their school. According to the teacher, "We did this because we hear so much about air quality here in the district, because our indoor environment affects all of us, teachers, students and administrators" (Unpublished Teacher Report, 1994, p. 1). Over the semester, the students and their teacher planned and implemented their study. They started their study by placing petri dishes near ventilation ducts for extended periods, then removing and culturing them. To their surprise, they discovered lots of molds growing in the petri dishes, but they could not quantify or measure the amounts across different locations within the school. Luckily for the students and their teacher, as members of the GLP, when they were asked to plan and implement a research project, they decided to focus on indoor air quality. According to Berenfeld (1994),

> Global Lab supported the students with TERC Air Pumps, a variety of air testing tubes, and instructional materials. After a series of tests, the students detected no

appreciable sulfur dioxide, ozone and carbon monoxide levels in their classrooms, but found consistently high carbon dioxide levels. They determined that over the course of the school day, CO_2 levels exceeded the recommended limits of 1000 parts per million (ppm) set by ASHRAE (American Society of Heating, Refrigerating and Air Conditioning Engineers). Mindful of the complaints about poor air throughout the school, the students conducted a schoolwide survey of air quality in other locations and sought the opinions of faculty. (p. 1)

The students discovered that in some locations in the school, the CO_2 levels were as high as 2,100 ppm. In addition, their survey of teachers found that a majority of teachers felt that the air quality at the school was bad. They reported symptoms such as headaches, sneezing, coughing, irritated eyes, dizziness, and other conditions. The students sought the advice of a scientist from the GLP, who told them that although the CO_2 levels were unusually high, the symptoms could not be determined without further investigation. Armed with their data, the students and their teacher presented their findings to school district authorities, who then visited the school and invited the local EPA officials to conduct their own test. The CO_2 levels arrived at were the same as the students' results. Because of this incident, the ventilation system in the school was repaired and the CO_2 levels dropped significantly.

Teacher's Role

The teacher's role in this project was to motivate the students to take on this project. She was able to convince her class that this project was something that would benefit them all. She also acted as team leader (principal investigator) and provided direction for the project design and data collection. She made sure that the project was aligned with school district and state learning standards. She divided the students into task-oriented groups and kept each group on task until they accomplished their task. She posed critical questions about the project at each phase so that the students could figure out what the next step was going to be.

Students' Role

The students' primary roles were to use tools and technology to collect data, analyze data, and conduct presentations. They used the air sampling pumps and test tubes to collect air samples during the first and last period of each day. When they collected sufficient data, they analyzed their air quality data, developed graphs, and compared their graphs across classrooms. They entered their data into the GLP database (and on the GLP online network). By comparing their data with other GLP schools doing similar projects, the students discovered that their results were unusual. With the guidance of their teacher, the students also developed a survey and distributed it to all the teachers in their school. Their survey consisted of ten questions on a five-point Likert scale. The students took responsibility for communicating with the authorities and their parents about the outcome of their research.

Strategies for Implementation

The implementation strategy used in this case could be described as teacher-student partnership. Here, the teacher and her students collaborated on the planning, implementing, and evaluating their project. The teacher provided leadership and engaged her students in the process of identifying the project, designing the study, developing the survey instrument, implementing the project, managing resources, facilitating the process, and mentoring the students.

Project Activities

Refer to the "Background and Context" section above.

Challenges and Lesson Learned

The challenge of implementing such a case is for teachers and their students to search and identify a topic that will generate sufficient interest and "investigative power" that will attract students' attention. This requires teachers interested in implementing PBSI to be proactive and be on the constant look out for ideas, especially in their own backyard. There are many lessons that could be learned from this case. However, according to Berenfeld,

> Students had been stimulated not by academic problems presented in textbooks, but by real-world problems that clearly affected their lives. For the educational process to be effective, students must want to learn. They are motivated to do so when they appreciate the relevance of the curriculum. What could be more relevant than examining the very air you breathe? (p. 20)

Case 3. Project-based Earth/Environmental Science: Park Ecology

Background and Context

City parks serve many purposes. For instance, they provide a space for city dwellers to play and enjoy the outdoors. Some parks have sport facilities, such as baseball fields, football fields, tennis courts, swimming pools, and soccer fields, that are available and open to the public to use. However, there are other uses of parks that are equally important—if not more important—to us. Parks are ecosystems that consist of plants and animals and the physical environment such as air, water, and soil. The interaction of all these organic and inorganic elements helps to replenish the oxygen we breathe and absorb the carbon dioxide we produce. Parks contribute to the complex process of energy and nutrient recycling that takes place in our natural environment. Without parks, life in our cities would be socially and environmentally unhealthy. Because our parks serve many purposes, it is important to maintain them and to monitor their health regularly.

In this case, students were asked to visit their neighborhood park and investigate its condition of health. A broad, ill-defined, guiding research question was provided as follows: How healthy is the park?

Teacher's Role

Prior to the implementation of this project, the teacher conducted a workshop for the students on how to describe a terrestrial ecosystem and an aquatic ecosystem. Students learned (1) how to collect baseline data on soils (testing and describing the characteristics of the soil at your park), (2) how to conduct soils tests for pH, nitrogen, phosphorus, and potassium, (3) how to collecting baseline data on water quality (testing and describing the characteristics of the water at your park), and (4) how to test for color, odor, pH, dissolve oxygen, nitrate, turbidity, and temperature. The students were also given a demonstration on how to use soil- and water-testing tools and equipment. Table 5.1 below shows the hardware, software, supplies, and materials that were provided to the students to support their projects.

Students were each given a handbook titled *Project-based Park Ecology*, which described the basic concepts and practices in ecology and environmental science. The expectations of PBSI, such as students taking responsibility

Table 5.1. List of Hardware, Software, Materials and Supplies Used in Park Ecology Project.

Hardware, Software, & Other Resources	Instructional Materials & Supplies	Laboratory/Fieldwork Materials & Supplies
• Laboratory space for 18 students • Microscopes • Computers with Microsoft Office & printers • TV monitor and VCR • Overhead projector • Data projector • Soil testing kit • Water testing kit • Digital cameras	• Student's research handbook • Notebooks • Pencils and pens • Poster paper • Markers • Rulers • Erasers	• Protective clothing* • Gloves & goggles • Test tubes (different sizes) • Test tube racks • Pipettes • Beakers (different sizes) • Freezer bags & glass bottles (medium and small sizes) • Buckets, hand trowels, & ropes • Measuring tape

*Students were required to put on appropriate outdoor clothing and shoes.

for their own learning, teachers serving as facilitators, mentors, and resource persons, and the importance of students' own questions as a driving force in PBSI were covered. The lifecycle of a project was reviewed and the role of the teachers and students at each stage discussed. The importance of collaboration in PBSI was emphasized, and the students were reminded that scientific investigations require collaboration in almost all tasks such as: proposal development, instrument development and testing, data collection and analysis, and reporting writing and presentation of findings.

Students' Role

The students divided themselves into two groups, one focusing on soil quality and plant life and the other on water quality. Each group conducted a rapid ecosystem evaluation (REE). A REE is a quick, systematic, and qualitative way of finding out how healthy an ecosystem is. It consists of two steps. The first step is to describe the ecosystem, and the second step is to collect simple baseline data to support and enhance the description. After completing their REE, each group identified a driving research question, conducted an in-depth investigation, and prepared a report. Finally, each group presented their reports for peer review at a public forum and invited their parents to attend.

Strategies for Implementation

The implementation strategy in this case could be described as part teacher-centered and part student-centered. The teacher set the stage, provided a broad definition of the project, conducted an orientation, provided training on procedures and use of tools and technology, and served as facilitator, resource, and mentor. The students then took full responsibility for planning, implementing, and evaluating their own projects, as well as their own learning.

Project Activities

Each group was asked to brainstorm and identify a driving question relating to the ecosystem of their local park that they would investigate. They shared their questions, which were written on a chalkboard. Each question was collectively reviewed using the following criteria: Is the question purposeful or going to serve a useful purpose? Can the question be investigated in the time allowed? Can the question be investigated with the resources available? Is the question new? If old, how can the question be restated to learn something new?

Each group was asked to prepare and submit a research plan. Students were asked to include a research question, procedure, tools and materials required, time required to complete the project, and roles and responsibilities. Students were also asked to include a statement on how they would assess and provide evidence of their own learning from their projects. Each group's research plan was reviewed and approved by the teacher-facilitator.

A pre assessment was administered to determine students' content knowledge, process skills, and dispositions as they relate to ecology and environmental science. Prior to the start of the session, all data collection materials and tools were provided. These included measuring tape, water-testing kits, soil-testing kits, digital cameras, hand trowels, buckets, ropes, gloves, goggles, notebook, pens, and pencils. Students reviewed safety issues, sampling procedures, and the importance of keeping accurate data.

Students were transported to a local park where they spent three hours collecting data. While collecting their data, the students were supervised, mentored, advised, assisted, and facilitated by a teacher with two support staff. Extra help was provided when necessary by the teacher and support staff. Group interaction, dialogue, and sharing of ideas were encouraged. At the end of the data collection, students gathered their materials and cleaned and packed all tools before they were transported back to school.

Prior to data analysis, the basic methods of analyzing data were discussed. Students were provided with a simple chart showing types of data and possible ways of analyzing them. They were also given a template for report writing. This was followed by a brief review of the rules of scientific writing. Students were asked to analyze their data and prepare summary reports or posters. The protocol for presentation and criteria to be used for evaluating project reports were discussed. As they analyzed and interpreted their data, the students were mentored, advised, and assisted by a teacher-facilitator with staff. Each group was provided with a computer equipped with Microsoft Word and Excel to facilitate the preparation of their report.

Students presented their reports for peer review and invited their parents. After the presentations were completed, each group was asked to reflect on their research process. In addition, they were asked to reflect on what they knew prior to and after conducting their projects in terms of ecology and environmental science concepts, science process skills, and dispositions. The students reflected and discussed areas for improvement and possible next steps. Finally, a post assessment to determine understanding and any gains in science concepts and process skills was administered.

Challenges and Lesson Learned

Although the implementation of the park ecology project was a success, it wasn't without challenges. For instance, during data collection in the field, it was found that there weren't enough hand trowels. As a result, additional trowels were immediately purchased from a nearby hardware store and brought to the site. While waiting for the trowels, some groups improvised techniques to collect their soil samples, which could have affected their sampling process and/or samples. Another challenge was that there were only two water-testing and soil-testing kits. As a result, groups had to take turns to use the kits and this caused delays and frustrations among some students. Once students returned to the laboratory, some groups wanted to conduct a quick Internet search, but the laboratory was located in an older building that did not have Wi-Fi Internet access. In addition, printing of participants' reports was handled in separate room, which also caused some delays. One of the groups collected more samples than they could analyze, and had to complete their data analysis after the project was over.

It is important to note that throughout the implementation of the unit, there was support staff that provided assistance to the teacher-facilitator. Without this support staff, it would be very difficult to implement the project and complete it in two days. One common reason given by some science teachers for not implementing PBSI is that it takes too much time. The implementation of this project demonstrated that it is possible for teachers and their students to implement PBSI even on a short-term basis.

Case 4. Project-based Physics: High-Altitude Ballooning

Background and Context

On September 28, 2015, NASA Jet Propulsion Laboratory at California Institute of Technology posted on its website, "NASA Confirms Evidence That Liquid Water Flows on Today's Mars." Then the site went on to say:

> New findings from NASA's Mars Reconnaissance Orbiter (MRO) provide the strongest evidence yet that liquid water flows intermittently on present-day Mars. Using an imaging spectrometer on MRO, researchers detected signatures of hydrated minerals on slopes where mysterious streaks are seen on the Red Planet. These darkish streaks appear to ebb and flow over time. They darken and appear to flow down steep slopes during warm seasons, and then fade in cooler seasons. They appear in several locations on Mars when temperatures are above minus 10 degrees Fahrenheit (minus 23 Celsius), and disappear at colder times. (front page)

This news was greeted with great excitement, particularly in the science community. The sciences that made this discovery happen, space sciences and related disciplines, are well aligned to students' curiosity because they are full of challenges, unanswered questions, wonder, mysteries, stories, potential benefits, and catastrophes. Space sciences span several disciplines, which includes and are not limited to astronomy, astrophysics, astrobiology, astrochemistry, cosmology, planetary science, climatology, metrology, remote sensing, geology, aviation, aerospace engineering, robotics, and satellites. Although these disciplines may be seen as accessible only to a few, in fact, teachers and students could engage in space science projects that will help students develop better understanding and love for science. One of the biggest unanswered questions humans have wrestled with since our beginnings on planet Earth is, how did our universe begin? Ballooning was one of the earliest methods humans used to explore the properties of space. According to Beck-Winchatz and Bramble (2014):

> The Kármán line at an altitude of 100 km is generally regarded as the boundary between Earth's atmosphere and outer space. Even though high-altitude balloons do not reach this altitude, they pass through the Pfotzer maximum and most of the ozone layer in the stratosphere, and expose payloads to high doses of cosmic and ultraviolet radiation. Because payloads are not subjected to microgravity, balloon flights can serve as controls that allow researchers to discriminate between the effects of radiation and microgravity. Balloons are also by far the most affordable and easiest way to provide students with direct access to a space like environment for their experiments. Flight hardware is readily available from several vendors and the procedures for developing a flight system and executing flights is well documented. (p. 118)

The High-Altitude Ballooning (HAB) Project is an example of a project that could help students meet both state and national science learning standards, while at the same time engaging in authentic science learning. As Hike and Beck-Winchatz (2015) put it,

> Many research projects require scientists to collaborate with engineers to develop the technology they need for such experiments. Ballooning engages high school students in a similar interaction and thus aligns with the Next Generation Science Standards emphasis on science and engineering practices (NGSS Lead States, 2013). (p. 30)

Below is a brief description of the HAB Project.

In spring 2013, the teachers and students at Curie Metro High School in Chicago implemented a HAB project as part of a chemistry final project. Most of the students were of Hispanic descent and from low-income families (Hike & Beck-Winchatz). The purpose of the HAB project was to investigate the following driving questions: (1) How do different liquids cool and freeze when exposed to the low temperature in the upper atmosphere? (2) How does the ozone layer affect ultraviolet light intensity? and (3) Does the ideal gas law hold true in Earth's upper atmosphere? Working in their groups, the students designed and implemented 12 experiments. They reviewed literature, designed experiments, collected and presented data, evaluated their projects, and conducted oral presentations. The experiments covered a wide range of disciplines, such as biology, chemistry, physics, and astronomy. To learn more about the students' experimental designs, data collection methods and data collected, analysis techniques, tools and technologies used, and so forth, please refer to Hike and Beck-Winchatz.

Teacher's Role

Beck-Winchatz and Bramble note that the role of teachers in implementing HAB projects is to construct a plan that includes an organizational framework, timeline, and assessments. They propose that teachers use assessment strategies at different phases of the HAB project. For instance, in cases where the purpose of the activities is

> designed to generate or summarize ideas, such as brainstorming ideas for investigations and note taking during presentations, group discussions, and lectures, assessments can be informal, giving students full credit simply for completing them. ... For assignments that are critically important for the safety and success of the balloon mission, such as design and fabrication of payload containers and the development of pre-launch procedures, students should be required to revise their work based on instructor feedback until it is approved by the instructor. (p. 125)

Students' Role

There were five specific tasks that students needed to complete during the implementation of the HAB project. They are as follows: review of literature (two class periods), experimental design (three class periods), data collection and presentation (one school day), evaluation (one class period), and presentations (one class period). All periods lasted 52 minutes (Hike & Beck-Winchatz). Students were responsible for forming their own teams, identifying their own research questions, designing their own experiments, collecting and reporting their data, and evaluating their own learning.

Strategies for Implementation

From the information available, we could infer that the implementation strategy used for the HAB project was a student-centered strategy. Students took full responsibility for all aspects of the HAB project, while their teachers serve as facilitators, mentors, and resource persons.

Project Activities

The project activities could be summarized as follows:

1. Formation of research teams
2. Braining storming, identifying, selecting, and defining of research questions

3. Conducting a review of the literature
4. Planning of experiments
5. Constructing of payloads, review of safety protocols
6. Launching, data collection, tracking, and retrieval of balloons
7. Assessment and presentations

Challenges and Lesson Learned

A major challenge of HAB is that it requires open space where there is less population density and trees to launch balloons. Because HAB requires the use of material, tools, and technology, funding must be available to purchase these items. According to Hike and Beck-Winchatz, the total cost for a typical flight system was estimated to be $1,480. In addition, some federal and state safety regulations govern ballooning, and compliance is required. Nevertheless, there is federal government funding from NASA to support partnerships among teachers-students-scientists to conduct HAB projects. HAB projects also have the potential benefits of promoting citizen science and scientific literacy among students.

Chapter Summary

In this chapter, we presented four cases of PBSI, namely, the herpetology project, carbon dioxide and indoor air quality project, park ecology project, and high-altitude ballooning project. What all these projects share in common is that they are student centered either in part or in whole. This means that students are in charge of their own learning, while their teachers take a back seat, monitoring, facilitating, mentoring, and supporting them throughout the learning process. Each project requires teachers and students to assume specific roles, demand specific teaching and learning activities, employ specific implementation strategies, and come with their unique challenges and opportunities for science learning. These projects capture a range of projects types and contexts and their applications to any new situation will be determined by the objective conditions prevailing at the particular school: school leadership, school culture, resources available, teachers and students' interest, knowledge, and dispositions.

Food for Thought

The following are questions that will help the reader reflect on the chapter.

1. Identify and describe four roles that teachers must play in a PBSI environment.
2. Identify and describe four roles that students must play in a PBSI environment.
3. What do you consider as the most important challenge that the four cases share in common?
4. Select one of the cases and describe how you would implement it differently.

· 6 ·

EVALUATING PROJECT-BASED SCIENCE INSTRUCTION

Chapter Overview

The evaluation of students' science learning in a project-based science environment presents a number of challenges for science educators. According to Kanis (1991), "Our most pervasive goal as science educators is to make students effective problem solvers. There appears to be a paradox, however, between this goal and the means by which we assess our students" (p. 290). The purpose of this chapter is to identify the methods currently used to assess students' science learning in project-based science environments and to critically examine their technical, curricula, economic, and social strengths and weaknesses.

Methods Currently Used to Assess Students in Project-based Science Learning Environments

In order to identify the methods currently used to assess students in project-based science learning environments, a preliminary search was conducted of the ERIC database (1961–1981 and 1982/1983–1994). Entries under the following headings were examined: assessment, performance-based assessment, alternative assessment, testing, evaluating, science process skills, student research skills, students' science attitudes, students' science misconceptions, project-based science, inquiry-oriented science, student-centered instruction, high school science, and secondary school science, to name a few. In addition, different combinations of the above search terms were tried out. A follow-up search was conducted, and this time the focus was from 1996 to the present. The following descriptors were used to refine and narrow the search: student projects, science instruction, inquiry, problem-based learning, evaluation methods, science tests, student evaluations, and performance-based assessment. In addition, the "Alternative Assessments in Practice Database" (Aschbacher, 1993) and the Northwest Regional Educational Laboratory (1994, 1995) database, "Improving Science and Mathematics Education: A Database and Catalog of Alternative Assessments," were searched. A Google search was conducted using some of the search terms and descriptors listed above to search and identify more recent research publications, articles, and reports pertaining to assessment of science learning in PBSI.

A review of the literature obtained from all these sources shows that there are several methods currently used to assess students' science learning in project-based learning environments. They include and are not limited to standardized multiple-choice tests, essays, observation, interviews, oral examinations, presentations, surveys, questionnaires, project reports, portfolios, performance-based assessment, concept mapping, film or video testing, drawings, computer-assisted tests, formative assessment, embedded assessment, and a combination of any of the above.

The review also shows that there were different varieties of PBSI and some of the names were used interchangeably. For instance, authors used the names *project-based science*, *project-based learning*, *project-based assessments*, *STEM project-based learning*, *problem-based learning*, *inquiry-based science*, or *inquiry-based science education*. The one thing that cuts across all these the methods of assessments discussed under these project-oriented, student-centered, active learning environments was the assessment of students' performance, what students are "doing" in their classrooms.

Table 6.1 provides a summary of the different methods of assessing science learning not only in PBSI but also in other similar learning environments, such as problem-based and inquiry-based science environments. As can be seen, different authors use or suggest different methods of assessing students' science performance in project-based or nontraditional, student-centered learning environments. It is very difficult to talk about different methods of assessing students' science learning in PBSI without first addressing the variability in methods of assessment and terms often used to describe them. For the purpose of clarity, we will distinguish among three common terms used sometimes interchangeably and confusingly, when people talk about assessment of students' science learning. The terms are *assessment, testing,* and *evaluation.* According to Wiggins (1993), "An assessment is a comprehensive, multifaceted analysis of performance; it must be judgment-based and personal. ... An educational test by contrast, is an 'instrument,' a measuring device. We construct an event to yield a measurement" (p. 13). Chittenden (1991) puts it in this way: "Test refers to a full range of devices from commercial instruments to teachers' own techniques—for checking up on students' learning" (p. 25). For instance, in the case of science assessment, tests could include quizzes, end-of-unit exams, and lab reports. The term *evaluation* means the process of determining value or worth of something. It is about making judgment about "what is meaningful" (Patton, 2015, p. 5). From these definitions, it is clear that assessment and evaluation have one thing in common and that is they are about making a judgment, while a test or testing is one among many tools used in the service of assessment of students' science learning.

The methods of assessing students in a project-based science environment could be classified according to the modes of response (Madaus, 1993; Black, 1987). Modes of response refers to the ways in which students answer assessment questions. According to Madaus there are only three ways in which students can respond to assessment questions:

1. Provide an oral or written answer or a product (e.g., essay, portfolio, or oral discourse).
2. Perform an act, which will then be scored against certain criteria (e.g., read aloud from a book, repair a carburetor, or perform an experiment).
3. Select an answer from among several options (e.g., multiple choice or true-false items).

Methods of assessing science learning in a project-based science environment will be discussed under these three broad categories: assessment by presentation or product, assessment by performance and observation, and assessment by choice selection. The focus of the discussion will be on students' science performance, and to capture the diversity of students' performance, the concept of project-based science learning environments will be broadened to include problem-based and inquiry-based science learning environments.

Table 6.1. Different Methods of Assessing Science Learning in Project-based, Problem-based, and/or Inquiry-based Science Environments.

Methods of Assessment	Author(s)
Quasi-experimental design (pretest, posttest, and control group)	Karaçalli & Korur (2014)
Interview protocol, observations, videos of student engagement, presentations, and worksheets	Avraamidou (2013)
Formative assessment (dialogic talk, teacher questions: open-ended vs. close-ended, subject-centered vs. person-centered, student self-assessment, peer assessment); summative assessment (performance assessment, embedded assessment, written items)	Harlen (2013)
Formative assessment (T-Chart, writing prompts, journal entries, and minute paper)	Trauth-Nare & Buck (2011)
Student laboratory reports, surveys, and focus group interviews	Davis, Lockwood-Cooke, & Hunt (2011)
Web-based portfolio assessment (use of experimental and control group)	Chang & Tseng (2011)
Questionnaire and in-depth interviews	Miedijensky & Tal (2009)
Observations and students' artifacts (use of pre- and posttests)	Rivet & Krajcik (2008)
Selected-response test, constructed-response test, performance assessment, portfolio assessment, and affective assessment	Popham (2007)
Pretest and posttest (multiple choice and free response items)	Marx, Blumenfeld, Krajcik, Fishman, Soloway, Geier, & Tal (2004)
Cognitive complexities of science assessments	Baxter & Glaser (1998)
Interview protocol and observations	Baxter, Elder, & Glaser (1994)
Selected response assessment, essay assessment, performance assessment, and personal communication (oral assessment)	Stiggins (1994)
Multiple choice, open-ended, and short answers, laboratory performance, tasks, portfolios	California Dept. of Education (1993)
Oral or written responses, product, performance, multiple choice, or true-false items	Madaus (1993)
Multiple choice, true-false, constructed response, essay, oral exams, exhibitions, experiments, portfolios	Office of Technology Assessment (1992)
Competence, motivation, and behavior	Raven (1992)
Hands-on performance assessment	Tetenbaum (1992); Shavelson, Baxter, & Pine (1991) Baxter, Shavelson, Goldman & Pine (1992)
Observations, performance assessment, tests	Chittenden (1991)

Methods of Assessment	Author(s)
Observations, verbal responses, written records, drawings, products	Hein (1991)
Group test, individual test, written test, oral test, speeded test, power test, pretest, posttest	Wiersma & Jurs (1990)
Standardized multiple-choice test	Mattheis & Nakayama (1988); Burns, Okey, & Wise (1985); Morgenstern & Renner (1984); Dillashaw & Okey (1980); Klopfer (1973)
General impressions, course work, assignments, pupil's self-assessment, rating scales, checklist, practical tests, written tests	Black (1987)
Paper and pencil tasks, demonstrations, computer-administered tasks, hands-on tasks, and various combinations	Educational Testing Service (1987)
Formal vs. informal assessment, formative vs. summative assessment, continuous vs. terminal assessment, course work vs. examinations, process vs. product assessment, internal vs. external assessment, convergent vs. divergent assessment, ideographic vs. nomothetic assessment	Rowntree (1987)
Formal assessment (standardized or norm-referenced tests) and informal assessment (teacher-made and criterion-referenced devices)	Evans, Evans, & Mercer (1986)
Clinical interview	Finley (1986)
Motion picture film	McIntyer (1972); Morgan (1971)
Computer-aided testing	Collins (1984)
Written reports, test items, laboratory practical exams, and observational assessment	Lunetta, Hofstein, & Giddings (1981)
Paper-and-pencil assessment, work-sampling assessment, observation, performance assessment, student-prepared assessment	Lien (1980)
Achievement tests, general mental ability tests, aptitude tests, interest inventories, attitude measures	Erickson & Wentling (1976)
Individual and group tests; classroom and standardized tests; oral, essay, and objective tests; speed, power, and mastery tests; verbal, nonverbal, and performance assessment; readiness and diagnostic tests; norm-referenced and criterion-referenced.	Ahmann & Glock (1971)
Oral tests, written tests, and performance assessment	Nedelsky (1965)
Oral recitation or oral quiz, observation of pupil performance, examination of written work	Lindvall (1961)

Assessment by Presentation and Product

The assessment practices included in this category are oral presentation, essay, project reports, drawings, and portfolios. Each method will be discussed in detail.

Oral Presentation

Records of assessment by oral presentation go back to the 1200s in Italy (Green, 1991; Rowntree, 1987), where, at the University of Bologna, professors conducted oral examinations to award degrees. Students were questioned by a panel of three examiners to assess their knowledge and understanding of their field of study. Two of the examiners were usually professors who were familiar with the student's work, while the third one was an expert on the subject matter, and may have come from outside the university. The scoring system for the oral examination was on a scale of 1 to 30, with each examiner allowed to assign points ranging from 1 to 10.

Oral examinations continue to be used in Italy, both at the university and secondary levels (Giordano, 1994). In the United States, oral examinations are used mostly at the university level. However, this method of assessment is also practiced in some high schools (Office of Technology Assessment, 1992). In order to assess student science learning under project-based curricula, science teachers and researchers frequently ask students to talk about the design process, procedures, and results of their work, either as individuals or in groups (Roseberry, Warren, & Conant, 1992; Carey, Evans, Honda, Jay, & Unger, 1989; Colley, 1997; Colley & Broderick, 1994). Some of these oral presentations may be formal or informal. They may also take place during a science fair. Oral presentations that take place during science fairs are sometimes accompanied with an exhibition of students' artifacts. Scoring is usually carried out by a panel of outside judges, using a scoring protocol or checklist of specific qualities deemed relevant to scientific understanding and practice.

Some researchers (Roseberry, Warren, & Conant; Guesne, 1992; Osborne & Freyberg, 1992; Carey et al., 1989; Nussbaum & Novak, 1976) interested in assessing students' conceptions about science and scientific inquiry use clinical interviews, which may be regarded as one variant of oral presentation. An example of clinical interview protocol is provided in figure 6.1. According to Patton (1987), a clinical interview "consists of a set of questions carefully worded and arranged for the purpose of taking each respondent through the same sequence" (p. 112). In carrying out clinical interviews, researchers develop core questions that probe students' epistemology about science and scientific inquiry. The data from clinical interviews are coded for specific categories and then analyzed.

SCIENCE CONTENT QUESTIONS
1. **Can you tell me all you know about the biological properties of water?**
The interviewer should look for responses that demonstrate students' knowledge about water as a medium for microorganisms. Students must also explain osmosis and diffusion of molecules as one of the primary mechanisms in which living cells take in and expel water, water enhances chemical reactions in living cells and regulates cellular temperature.

2. **Can you tell me all you know about the chemical properties of water?**
The interviewer should look for responses that demonstrate students' knowledge about composition of water—2 hydrogen and 1 oxygen; water is a polar molecule because the hydrogen atoms are negatively charged while the oxygen atoms are positively charged. This attractive bonding forms a weak bonding called hydrogen bonding—this is why water is adhesive and clings to surfaces; pH of water—pH is a measure of the hydrogen ion concentration in the water. Pure water has a neutral pH of 7.0 (pH scale ranges from 1 to 14, with 1 being very acidic and 14 being very basic).

3. **Can you tell me all you know about the physical properties of water?**
The interviewer should look for responses that demonstrate students' knowledge about water as an energy source—flow of water from dams can power turbines, which generate electricity; freezing, melting, and boiling points; evaporation and condensation; water pressure; buoyancy of water.

SCIENCE PROCESS SKILLS QUESTIONS
1. **Explain to me what you know about science.**
The interviewer should look for responses that includes science as having a variety of meanings based on the context in which it is used: science as a way of thinking and knowing; science as a collection of disciplinary knowledge; science as a systematic process of finding out; science as a social activity; and science as an industry.

2. **What do you think are the qualities of a scientist?**
The interviewer should look for responses that include love for science, love for the environment or nature, curiosity, thoughtful, observant, patient, persistence, imagination, creativity, intuition, open-minded, strong knowledge in a science content area, and training in scientific methods and techniques.

3. **Can you explain to me in a step-by-step way, how a scientist investigates a question or problem?**
The interviewer should look for students' ability to identify the main steps in the scientific research process, such as: formulating a research question/hypothesis, designing an experiment, collecting data, analyzing data, drawing a valid conclusion, generating theory and/or posing new questions. Students should also be able to distinguish between a "control" and a "treatment" in a control experiment. Responses should also demonstrate students' understanding of types of variables, and the difference between a random and nonrandom sample.

4. **Why do you think scientists collaborate on scientific investigations?**
The interviewer should look for possible responses such as: "a group working on a scientific investigation can cover more work easily compared to individuals; scientific investigation requires different tasks and skills and each individual in a group can use the skill they are good at and tackle a specific task; because group members can share ideas and learn from each other."

Source: Colley, 2001.

Figure 6.1. Clinical Interview Protocol from the tBLISS Project.

Clinical interviews can also be in the form of "think-aloud-problems" (Roseberry, Warren, & Conant). Think-aloud problems are short accounts of real or hypothetical scientific scenarios designed to assess science process skills. They are usually read aloud to students, who then think and discuss them before giving a response. These problems or scenarios cover a wide range of science process skills, such as problem solving, critical thinking, ability to conceptualize a research problem, identifying variables, interpreting data, and drawing conclusions. The "Boston Harbor Problem" and the "Sick Kid Problem" are typical examples of think-aloud problems (see figure 6.2). They were developed by Roseberry, Warren, and Conant to investigate the acquisition of science process skills by language minority students. The Boston Harbor Problem was selected to represent what the authors call "near transfer";

Problem 1: Boston Harbor
I'm going to tell you a true story; it's sort of a mystery. It's about the Boston Harbor. In the last few years, people have noticed that there is something wrong with the water in the Harbor but no one knows exactly what is wrong.
   Fishermen have noticed that there are fewer fish in the Harbor. And they have seen a lot more algae. People who spend time near the Harbor have noticed that the water looks dirty; it is brown and foamy. It also has garbage in it. Tin cans, paper, and old food float in the water. Sometimes you can even see dead fish floating on the waves.
   You are a famous scientist. The mayor of Boston asks you to find out what is wrong with the water.

- What is the first thing you do?
- What do you think might be wrong with the water?
- How will you find out if you are right?
- Do you have any ideas about how you could make the water clean again?

Problem 2: Sick Kid
I'm going to tell you another true story; it's a mystery, too. It's about some children in a school who get sick and, when it happened, no one knew what was making them sick.
   It happened in a town just outside Boston. All the children in an elementary school were watching a play put on by the sixth graders. Suddenly, a boy in the play fell off the stage and cut his chin. He said he felt sick and some teachers carried him to the nurse. Then a student watching the play got dizzy and fainted. Then some other students felt sick to their stomachs. Suddenly, lots of students were sick.
   You are a famous scientist and you live next door to the school. When the children get sick, the principal runs over to your house and asks you to come and find out what is making the children sick. You agree and go to the school.

- What is the first thing you do?
- What do you think might be making the children sick?
- How will you find out if you were right?

Source: Roseberry, Warren & Conant, -1992.

Figure 6.2. Think-Aloud Problems.

it required students to think through a problem they encountered in class involving water contamination. They are then asked to apply the knowledge they acquired in the context of the water quality investigations. The Sick Kid Problem represents "far transfer" and requires students to reason through a problem they have not previously studied and apply their scientific theories to solve it. Think-aloud problems often end with core questions, but they may be modified to yield a truer picture of students' scientific epistemology.

The scoring of clinical interviews and think-aloud problems varies greatly, and as a result it is difficult to devise a way for estimating their reliability. In most cases, predetermined scoring protocols are developed and tested. A panel of experts or group of teachers is then recruited to score students' responses. Disagreement among panelists are then discussed and a consensus on the final score is reached (Carey et al., 1989; Baxter, Shavelson, Goldman, & Pine, 1992).

Strengths of Oral Presentation

Technical. Oral presentation can be conducted in a variety of ways, such as oral examination, clinical interview, informal conversation, and group discussion. Because of its many forms, oral presentation is very flexible and can be administered in an individual as well as a group setting. It is particularly useful if the number of students involved is small. Immediate feedback is possible between assessment and instruction (Stiggins, 1995). The oral presentation can be transformed into a written form (transcript), coded, analyzed, and scored. Scoring can be done by the student's teacher, as well as by an outside judge. Interscorer reliability can, therefore, be estimated. The time required to prepare oral presentations is relatively short, thus allowing extra time for other curricula activities.

Curricula. Verbal responses provide descriptions of students' misconceptions, explanations, predictions, and understanding of science. According to Hein (1991), "Verbal responses are a particularly useful way of finding out what students know, since they make up much of the day-to-day interchange between teachers and pupils" (p. 111). They are suitable for assessing students' reasoning, attitudes, values, interests, and problem-solving abilities. By providing students with the opportunity to respond verbally or give oral presentations before and after a science unit is taught, teachers can assess students' baseline knowledge and how such knowledge has changed over time (Hein, 1991). Oral presentation can provide direct information on student learning styles and therefore help teachers to diversify their teaching strategies. In addition, it can help students develop better communication skills, by learning the rules and actually taking part in the process of oral discourse. An assessor can ask a follow-up question to probe deeply students' answers or take advantage of responses that could be used to enhance the assessment process.

Social. The social strength of the oral presentation lies in the fact that it can be conducted in more than one language. This means that it can cut across ethnicity or cultural boundaries. Some models of oral presentation, such as "science talk" (Theberge, Morrison, & Crowder, 1993), empower students by providing them with the opportunity to respond to a question in more than one way (storytelling, correct answers, and sense making). Such an approach takes into consideration the sociolinguistic background of each student. When done right, oral presentation can create a sense of community and empower both teachers and students.

Weaknesses of Oral Presentation

Technical. The use of oral presentation as a method of assessment assumes that all students are sufficiently prepared in the art of communication. In reality, some students lack these skills and are, therefore, at a disadvantage when confronted with oral assessment. There is a high degree of subjectivity and distortion in the recording and scoring of oral information (Patton, 1990). According to Stiggins (1994), this is because of the potential "problem of forgetting," the personal and professional biases of the assessors, as well as the quantity and quality of the questions asked. Elaborating on the above factors, Stiggins (1994) notes the following:

> We must understand the fallibility of the human mind as a recording device. Not only can we lose things in there, but the things we put in can change over time for various reasons, only some of which are within our control. ... We must also remain aware of and strive to understand those personal and professional filters, developed over years of experience, through which we hear and process student responses. They represent norms or standards, if you will, that allow us to interpret and act upon the achievement information that comes to us through observation and personal communication. (p. 212)

Because some students talk too fast, teachers or assessors sometimes use tape recorders to record student responses. There are a number of problems associated with the use of such technology. One such problem is that it may influence students' responses by making them overcautious or overexcited about their voice being recorded. In addition, it may take attention away from the subject matter during an interview. Tape recorders are not perfect and are subject to operational failures during the process of assessment.

Some students are particularly good at learning how to respond in a politically correct way. In discussing this particular problem, Bell, Osborn, and Tasker (1992) note,

> Children spend a considerable portion of their childhood learning how to please their elders, and they are adept at fastening on small cues as to what is expected of them. In a teaching role, which even non teachers adopt from time to time, we are prone to use leading questions, to reject "wrong" answers by raising our eyebrows or rephrasing the question, and to praise the right answer when we get it. (p. 151)

Curricula. Oral presentations are less effective in determining students' psychomotor skills (Stiggins, 1994). One can make references about students' abilities to perform assessment tasks from their oral responses; however, this is not very reliable, since what students may say they can do may differ from their actual practice.

Social. One of the social weakness of assessment by oral presentation is that it assumes all students can communicate in a particular medium without difficulty. On the contrary, language minority students are present in most American classrooms, and they often encounter difficulty in using English as a second language (ESL). In their study of how language minority students learn science, Rosebery, Warren, and Conant found that true science learning for this group of students can occur only when they are given the opportunity to engage in scientific discourse using a language in which they are proficient. According to the researchers,

> Typically, learning science in language minority classrooms (when it is taught at all) means learning English in the context of science content. Students memorize the definition of the word "hypothesis" but never experience what it means to formulate one. The emphasis is squarely on learning English vocabulary and grammar, with science as one means to that end. (p. 1)

Another assumption of assessment by oral presentation is that all students come from cultures that emphasize self-expression. The truth of the matter is that some students come from cultures that de-emphasize self-expressive behaviors. It would be unfair to assess all students using only oral methods without a thorough understanding of the students' cultural backgrounds. If oral assessment is to be used, the assessor and the assessee must share a common language. "You must know how to make meaning in the language and culture of your students. When you lack that understanding, mismeasurement is assured" (Stiggins, 1994, p. 218).

Economic. The administration and scoring of an oral presentation requires time and money. Time is required to implement student interviews, to transcribe student responses, as well as to read and score them. A conservative estimate of the time it takes to transcribe a one-hour audio tape is between 4 to 6 hours, although the time may vary, at an estimated cost of $16 to $18 an hour (MacTemporary Services, 1993). If outside personnel are hired to carry out oral assessment, then costs could easily run high.

Essay

Although assessment by essay has been practiced in different parts of the world at different times, the use of this form of assessment was first reported in China about 587 B.C. (Miyazaki, 1976). In order to admit people into the civil service, Chinese emperors held essay examinations in civil law, military affairs, agriculture, geography, classic archery, horsemanship, poetry, and music. The nature and format of these so-called eight-legged essays were very specific. Examinees were required to include a certain number of calligraphic characters and eight paragraphs arranged as follows: analysis of the theme, amplification of theme's explanation, past-explanation, argument 1, reassertion of theme, argument 2, and argument 3 (Rowntree).

In 1845, Horace Mann, superintendent of Boston Public Schools, introduced the first written essay examination in the American school system (Madaus; Morris, 1969). Although Mann had some political motives—accountability and control of Boston headmasters—for introducing written essay examinations (Madaus; Office of Technology Assessment), he cited the following reasons for his decision:

1. Written essay examinations were impartial because all candidates were measured by the same standards.
2. They gave students an opportunity to express themselves.
3. They allowed more questions to be posed as opposed to a time constraint oral examination.
4. They minimized teachers' interference.
5. They allowed students to demonstrate relationships between facts.
6. They provided transcripts of students' work.

The use of essays to assess student science process skills has been well documented by the state of California (California Department of Education). This type of assessment "challenged students to apply qualitative and quantitative information to new situations, design experiments, and use creativity in problem solving" (p. 8). It requires students to respond to open-ended questions or problems in writing, sometimes in limited time frames. Essays are constructed so that they are clearly understood by the students. In order to ensure clarity, directional words or prompts such as "explain," "interpret," "describe," "hypothesize," "analyze," "show," "relate," "compare," "contrast," and so on are used. The main assumption behind assessment by essay is that students must be able to read and understand and also can write well. In an essay examination or test, the student is free to construct responses, but in some instances there may be word limits.

The scoring of essays varies, but in general there are three basic procedures. The first is known as global or holistic scoring and suggests that the student's essay be evaluated as a whole piece. The second essay scoring procedure is called Objective Tests of Essay Answers (OTEA) (Nedelsky, 1965). The assumption behind this procedure is that the student, more than anyone else, understands the meaning of his or her essay best. The third essay scoring procedure is the opposite of the holistic procedure and is often referred to as the point-method (Stiggins, 1994). It involves breaking down the student's essay into specific themes or categories and assigning points for each category.

In a holistic scoring system, the assessor scores the essay on a rubric or scale. For instance, in the California Golden State Examination, a student's essay is scored on a scale of 1 to 6 (California Department of Education, 1993). Here, 1 means that the student demonstrated work of low caliber; does not show how facts are related; does not show "far transfer" across science disciplines; does not support facts or statements made; does not use tables, figures, and mathematical calculations to support his or her essay; does not apply scientific concepts to social, ethical, and environmental issues; and does not communicate well in written language (California Department of Education). A score of 6 means that the students have demonstrated works of high caliber. They are able to demonstrate the interrelationship of facts; are able to show "far transfer" across science disciplines; are able to use theories, models, and principles to explain scientific events or problems; are able to integrate text with tables, figures, and mathematical calculations; are able to apply scientific concepts to social, ethical, and environmental issues; and can communicate well in written language (California Department of Education).

The OTEA procedure includes the following steps. The students write an essay and then use their essays to answer standardized multiple-choice, objective tests, which cover the theme discussed in their essays. The scores from their tests are the determining scores for their essays.

The point-method of scoring is used by most science teachers because it is less time consuming. It requires teachers or assessors to establish specific criteria for assessment prior to the essay examination. Each criterion may be allocated a point or points. The essay is then read and comments made on the edges or sides of the essays. It is then scored according to the set criteria and the score converted into a percentage. The point-method is most suitable in scoring essays that are structured or semi open-ended. One of the main shortfalls of this procedure is that it pressures students to conform to a certain essay structure. In addition, it does not allow the assessors the flexibility to give credits to individual creativity. An example of an essay examination used in the assessment of science process skills is shown in figure 6.3.

INSTRUCTION

Answer any TWO of the following questions below. Each essay question is worth 20 points. Points will be distributed based on the following: (a) adequate depth and breadth of content, (b) coherence and clarity of written responses to the question, (c) organization and presentation of responses, and (d) originality and creativity.

1. Think of a question you want to investigate and explain how you would design an experiment to answer the question. In your response, identify and list all your materials and tools, describe your procedure and explain your independent and dependent variable.

2. In the past decade, we have witnessed an increase in the incidence of pandemic diseases such as SARS (severe acute respiratory syndrome), H1N1 (Swine Flu), Ebola, and Zika? Discuss the major factors responsible for the occurrence and spread of these diseases?

3. A class observes two demonstrations: water changing into steam and a piece of wood burning and producing smoke. A student concludes that both demonstrations must be examples of a chemical change because a gas is produced in each. Is the student's conclusion accurate? Explain your answer, referring to both demonstrations.

4. Two farmers notice that some bean plants are much taller than others, even though they are growing in the same field. One farmer thinks the difference in height is due to inheritance. The other farmer thinks it is because some plants in the field get more water than others. Describe an experiment that will provide evidence for which farmer is right.

5. Emissions of greenhouse gases from automobiles and factories are often cited as a cause of global warming. Automobiles and factories also emit solid particles such as smoke and ash.
Explain how emissions of these solid particles could cause global cooling.

Source for Questions 3–5: U.S. Department of Education, Institute of Education Sciences, National Center for Education Statistics, National Assessment of Educational Progress (NAEP), 2009 and 2011 Science Assessment.

Figure 6.3. Essay Examination Used to Assess Science Process Skills.

Strengths of Essay Assessment

Technical. The technical strength of essay assessment relies on the fact that the time required to prepare and administer them is short. Furthermore, they can be used for both small and large numbers of students. Students have more time to organize and present their ideas in an essay assessment. Methods of scoring essays are flexible and allow the assessor or scorer to use his or her own judgment sometimes.

Curricula. Essays are particularly suitable for the assessment of students' abilities to "organize, integrate and express ideas" (Gronlund, 1993, p. 84). In situations where it is not feasible to conduct a performance task, an essay can be administered. This will require students to enumerate and explain all the steps involved in performing a particular task. If the students explain all the steps without omitting any, it could be assumed that they have the skills required for the task.

Assessment by essay provides students with the freedom to respond in their own way, without having to select from a list of options. In addition, it helps students to improve their writing skills. Although the evidence on the influence of essay preparation on learning outcomes is inconclusive (Lundeberg & Fox, 1991), experience has shown that most students spend more time studying for essay assessment than for other types of assessment, so, as a result, they get more exposure to the subject matter. After they are scored, essays are sometimes used to enhance instruction. For example, it is common practice for teachers to reserve exemplary essays as models for other students to study. Some teachers often read aloud exemplary essays and use them to generate discussion.

Social. Essays, like oral presentations, provide opportunities for students to express themselves in their own words. In this way, students bring their own experiences into the classroom. Essays are a good way of encouraging debate on issues of social concern (e.g., crime, healthcare, and environmental pollution). Warren (1979) tested students' science content knowledge using essay versus multiple-choice tests, and found that the essay test was a better predictor of students' knowledge. In addition, most students who participated in his study felt that essay tests were better for testing their knowledge.

Economic. Essay assessment saves money in preparation and in administration costs because it requires less time. When the stakes are not high and the number of students to be assessed is relatively small, students can be used to score their own essay assessments using a holistic scoring rubric. This can help save money.

Weaknesses of Essay Assessment

Technical. The use of essay assessment assumes that all students can express themselves well in writing, when in fact there are vast numbers of students who cannot communicate well in writing. In an essay assessment, therefore, students' scores are influenced by the quality of their handwriting (good versus poor handwriting) and their adherence to the tenets of good grammar (how well they spell, punctuate, use tenses, style, etc.). Scoring is also affected by the individual personal and professional biases of the assessor. For instance, an assessor who believes in the right to choose will most likely score favorably a student's essay arguing for the right to choose than an assessor who does not hold any opinion on the issue whatsoever. Essay assessments have a low reliability not only because of the variation in how assessors assign scores, but also because there are fewer items to which the student can respond (Stiggins, 1994; Nedelsky, 1965).

Curricula. As a measure of overall course objectives, essay assessments are not suitable because they contain only a few assessment items or samples. The curricular weaknesses of oral presentations also apply to essay assessments.

Social. Because the procedures for scoring essay assessments are based on subjective judgment, essay assessments give teachers more authority over students' scores. Consequently, students who express views of dissent may receive lower scores.

Economic. Essay assessments are easy to prepare, but tedious to score. This is particularly true if they are used on a large scale. According to the Office of Technology Assessment:

> In general, the more open-ended a test is, the more expensive it will be to score, since scoring requires labor-intensive human judgment as opposed to machine scoring. ... For example, written examinations taken at age 16+ in Great Britain and Ireland cost roughly $110 per student. (In Ireland, candidates pay about 40 percent of the cost.) These costs may be tolerable in countries where a small percentage of the age cohort takes the examination. But in the United States, with nearly five times as many students in this age group, testing the 3,000,000 16-year olds in U.S. schools using the British or Irish model would cost about $330 million. Looked at from the perspective of one state, Massachusetts, it would cost almost $7 million to test all 65,000 16-year-old-students using the model of essay on demand; at present, Massachusetts spends just $1.2 million to test reading, writing, and arithmetic achievements of students in three grades and three subjects. (p. 141)

Project Reports

The idea of using student reports as an indicator of science learning is not new. For a historical perspective on the project approach to learning, please refer to Kliebard (1986), Stevenson (1928), and Stockton (1920). Since the introduction of project-based science curricula, students have been required to document their competence in carrying out science research projects. By carrying out research projects, students define problems, formulate hypotheses, design experiments, collect data, analyze data, and draw conclusions. In order to assess whether or not students have acquired these science process skills, teachers ask students to prepare reports of their work. The use of reports as a form of assessment assumes that students can articulate their ideas in words and that they are familiar with the guidelines for report writing.

There are several varieties of reports that are used in assessment of process skills (Greig, Wise, & Lomask, 1994; California Department of Education; Lunetta, Hofstein, & Giddings, 1981). Some follow the traditional format that one finds in scientific journals (i.e., abstract, introduction, materials and methods, results, and conclusion). Others follow a more flexible format that allows the student to use his or her own format (i.e., introduction to the project, what worked, what did not work, reflections). The scoring of project reports varies. For instance, when a group of science teachers were interviewed about what criteria they would use to evaluate student research reports (Colley, 1993), some teachers suggested that they would look for evidence of collaboration and critical reflection. Others said that they would look for evidence of clear formulation of research questions, calibration of instruments, and use of instruments. There are some who felt that the ability to communicate, as well as awareness of the social and environmental implications of their research project, was important.

Drawings

Drawing is another form of communication and expression. In science education, drawings are used to show the relationships between structure and functions. In biology, for example, drawings of a human skeleton are often used to teach students the different bones of the human body and how they work. Drawings are also an aid to the observation of scientific objects, events, and facts. According to Gallentine (1968), there are two types of drawings used in science, namely representative and diagrammatical (or analytical). In the former, the objective is to draw the object under study as close as possible to reality, both in size and form. In the latter, the objective is to show the different parts of the object and how they relate to each other. In order to do an analytical drawing, one must first analyze the object.

The use of drawings as a tool for assessing science process skills has been discussed by Hein (1991), who notes: "In my own work, I found that teachers could test students' knowledge of how to use a microscope simply by asking them to draw what they saw through the microscope" (p. 144). Other than asking students to respond to assessment questions by drawing, some researchers interested in students' misconceptions in science use prepared drawings and ask students to respond to them. For instance, to test whether children 10–11 years old were able to place a shadow, an object, and a source of light in the correct relationship to one another, Guesne presented a drawing that contained the sun, an object, and several possible directions in which the object's shadow may fall. The students were asked to determine the direction of the shadow if the sun was in front of the object and behind the object. In addition, students were asked to circle the right direction on another drawing.

The methods of scoring drawings vary; however, in many cases they are scored using a holistic approach or the point-method.

Strengths and Weaknesses of Assessment Using Project Report and Drawings

Report preparation and drawings are a form of communication that also involve writing, or the interaction of mind, pencil, and paper. This means that they are similar, if not the same, as assessment by oral presentation or essay. The same assumptions, strengths, and weaknesses of assessment using essay, as well as oral presentation, naturally apply.

Portfolios

The definition of portfolios varies. For instance, the Office of Technology Assessment refers to portfolios as "files or folders that contain collections of a student's work. They furnish a broad portrait of individual performance, assembled over time." Wolf (1988) notes that the term *portfolio* means "a chronologically sequenced collection of work that records the longer term evolution (the macrogenesis) of artistic thinking" (p. 27). Most portfolios, according to Wolf, contain a wide variety of students' work, as well as evidence of independent work. Paulson, Paulson, and Meyer (1991) define a portfolio as "a purposeful collection of student work that exhibits the student's efforts, progress, and achievements in one or more areas. The collection must include student participation in selecting contents, the criteria for selection, the criteria for judging merit, and evidence of student self-reflection" (p. 60).

By examining the above definitions of portfolio closely, the reader will notice that they all share three common characteristics. These include the following: (1) a collection of students' work, (2) accumulated or evolved over time, and (3) containing a wide variety or representative samples. Based on these three characteristics, one can say that the above definitions of portfolio are all acceptable. However, for the purpose of this paper, Paulson et al.'s definition will be used because, in addition to the three main characteristics of a portfolio, it includes criteria for selecting a portfolio, criteria for judging merit, and student self-reflection. These are critical to the development of any portfolios for assessment in any field.

The use of portfolios in assessment is more common in writing and the language arts than in science (Office of Technology Assessment). However, some states, such as California and Vermont, have experimented with portfolios as a tool for assessing students' science process skills. The experiences of California in the use of portfolios in science assessment are inconclusive because this state piloted its first portfolio assessment in science in 1992–1993. Nevertheless, the case of Vermont is very instructive, even though it focuses on writing and mathematical skills in the fourth and eighth grades. Three reasons why the Vermont portfolio assessment warrants a discussion here are (1) because it is an excellent example of a large-scale portfolio assessment implemented with some degree of success, (2) because the student portfolios assess process skills that are essential in science learning, and (3) because the steps taken to design and implement the portfolio assessment are applicable to science education.

The Vermont Portfolio Assessment Project (VPAP) was born in the late 1980s due to the general feeling, on the part of parents, teachers, and state education policymakers, that standardized tests were not an accurate indicator of student performance and school progress (Koretz, Stecher, & Deibert, 1992; Abruscato, 1993). As a result of the weakness in the current system of assessment, state education policymakers initiated a debate about assessment of students and school accountability. The debate and public discussion that followed led to the creation and funding of the VPAP. The VPAP was piloted in writing and mathematics for fourth and eighth graders in 1990–1991 in 137 schools. In 1991–192, the Vermont State Department of Education requested portfolios to be developed for assessment. Portfolio assessors or evaluators were recruited and trained. A special committee called the Vermont Portfolio

Assessment Benchmark Committee was formed to provide guidance and develop criteria for judging student work. The members of the committee included teachers, state department of education officials, and others.

The findings from the VPAP pilot showed that most students in Vermont could write well (Abruscato). Only 47 percent of fourth graders and 52 percent of eighth graders showed evidence of problem-solving skills. On the subject of mathematical communication, only 17 percent of fourth graders and 6 percent of eighth graders showed evidence of competency (Abruscato). There are many lessons that science educators and assessors can learn from the VPAP—in particular, the implementation process and the problems encountered.

The designing, implementing, and evaluating of portfolio assessment in science classrooms has been discussed in a number of articles (Collins, 1992a, 1992b). The first step in the process is deciding what needs to be included in the portfolio—in other words, developing a criterion for the selection of materials that will go into the portfolio. This is usually done by the teacher, but it could also be carried out by both teacher and student. The second step is deciding on criteria for judging students' work. These should be made explicit to the student so that he or she will know exactly what is expected. According to Stiggins (2007):

> Assessment for learning begins when teachers share achievement targets with students, presenting those expectations in student-friendly language accompanied by examples of exemplary student work. Then, frequent self-assessments provide students (and teachers) with continual access to descriptive feedback in amounts they can manage effectively without being overwhelmed. Thus, students can chart their trajectory toward the transparent achievement targets their teachers have established. The students' role is to strive to understand what success looks like, to use feedback from each assessment to discover where they are now in relation to where they want to be, and to determine how to do better the next time. (p. 22)

The third step is setting up a timetable for the completion of the portfolio. This will vary, depending on purpose, context, subject matter, and resources. The fourth step is very practical and involves collecting, assembling, sorting, organizing, selecting, cataloging, and documenting the work that will constitute the portfolio. At this stage, it is important to make use of available tools that can enhance the portfolio, such as computer word processing, audio and video records, CD-ROM, models, and photographs. These tools empower students and create a sense of control over their own learning (Stiggins, 1994). The final step in the process is the evaluation of the portfolio. This involves the teacher and students revisiting the criteria for selection and judging student work and discussing the portfolio. At this moment, both teacher and student engage in what Howard Gardner refers to as "perception and reflection" (Brandt, 1987). According to Gardner, "Perception means learning to see better, to hear better, to make finer discriminations, to see connections between things. Reflection means to be able to step back from both your production (work) and your perceptions and say, 'What am I doing? Why am I doing it? What am I learning? What am I trying to achieve? Am I being successful? How can I revise my performance in a desirable way?'" (Brandt, p. 32). By engaging in perception and reflection, the teacher and student share responsibility for the learning situation. In addition, the student discovers his or her own strengths and weaknesses and knows where to focus. The teacher is able to see a real picture of his or her student, as well as the effects of his or her actions.

Strengths of Portfolio Assessment

Technical. Portfolios serve different functions and allow great flexibility in design, content, and scoring. For instance, a portfolio can be used to evaluate students' achievement, determine instructional impact, or to encourage collaborative learning between teachers and students.

Curricula. In the context of the assessment of science process skills, portfolios are appropriate for assessing students' ability to design experiments, interpret results, and think critically. The curricular strength of portfolio assessment has been discussed by several authors (Wolf, 1988, 1989; Koretz, Stecher, & Deibert; Frazier & Paulson, 1992; Collins, 1992b); however, Paulson, Paulson, and Meyer have been able to articulate the advantages best:

> Portfolios have the potential to reveal a lot about their creators. They can become a window into the students' heads, a means for both staff and students to understand the educational process at the level of the individual learner. They can be powerful educational tools for encouraging students to take charge of their own learning. (p. 60)

Social. Students who speak English as a second language and those from other cultures are often at a disadvantage when assessed using methods that require proficiency in both written and spoken English. However, with portfolio assessment, they become empowered and are able to demonstrate their true knowledge, skills, and creativity (Wolf, 1988). In my own work, I implemented portfolio assessment in a general biology class and when I surveyed students about their impressions of the assessment methods used in the course, they overwhelmingly stated that they liked portfolio assessment because it was less stressful compared to standardized tests, and it helped them understand the material better.

Economic. If portfolios are designed as a joint venture between the teacher and his or her students, then the labor cost can be very low. This is because both parties collaborate in deciding the contents and scoring criteria, as well as participating in judging the final product through "peer grading" (Knight, 1992).

Weaknesses of Portfolio Assessment

Technical. If portfolio assessment is to be implemented statewide, as in the case of Vermont, the designing, administration, and judging of the portfolios can pose a major challenge. In their study of the VPAP, Koretz, Stecher, and Deibert note that more than 80 percent of fourth-grade teachers and more than 60 percent of eighth-grade teachers had difficulty covering the required curriculum. In addition, 60 percent of both groups did not prepare portfolio lessons due to time constraints. Without well-defined scoring criteria, the judging of students' work is bound to be subjective. To use students as peer graders, teachers must give coaching to students as to how to judge other peoples work without hurting their feelings.

Curricula. Portfolio assessment can interfere with school curricula by taking time away from other course work and by devoting it to producing, resubmitting student work, and conducting oral presentations. In schools where students have to cover a certain amount of materials in each course in order to be promoted to the next grade level, this may create conflict and imbalance in student learning.

Social. Introducing portfolio assessment into the high school curriculum, when most American colleges still require standardized test scores for admission, may create a conflict. Many teachers, students, administrators, and parents are concerned about this conflict, and as a result have become very cautious and hesitant about using portfolios. According to Wolf, LeMahieu, and Eresh (1992):

> Parents worry that portfolio work will not prepare their children for the "real world" of high school, college entrance exams and other tests. There is a question about whether receiving high schools will honor portfolios when they are accustomed to transcripts and tests scores. Staff in testing and evaluation know that alternative forms of assessment are in their infancy. And some have real doubts. What do we really know about validity? About gender and racial equity in portfolios? (p. 13)

The demands of portfolio work could also be a burden for students from ethnic minorities, instead of an opportunity to show what they have learned. As Hill and Larsen (1992) point out, students from ethnic minority groups are often overwhelmed because they do not have adequate resources at home, such as a computer, access to reference materials, ample space, time, and persons willing to offer feedback, with which to respond to the demands of portfolio construction. In addition, students may learn new ideas and strategies that are appropriate for their communities only to find out that they do not fit the traditional academic norm.

Economic. Implementing portfolios assessment on a large scale is costly in terms of teacher, as well as class, time spent on portfolios activities. In their study of the VPAP, Koretz, Stecher, and Deibert surveyed 172 teachers (124 fourth grade and 48 eighth grade) about the time they devoted to portfolios. The researchers found that teachers spent a lot of time on portfolio activities both in and out of class. On average, teachers spent 30.1 hours of their own time and 13.7 hours of class time on portfolio activity each month. Preparing for and conducting portfolio lessons consumed most of the teachers' time, while doing portfolios for the first time was reported as taking most of the class time. Other than time, an initial training of teachers, administrators, and portfolio raters is required if portfolio assessment is to be implemented. A school that is located in a poor or disadvantaged district may not be able to implement portfolio assessment due to the costs associated with teacher time and training.

Assessment by Performance and Observation

A distinction needs to be made between the headings or phrases *assessment by performance and observation* and *performance assessment*. In the context of this paper, assessment by performance and observation refers to assessment methods that require students to perform a task under observation. The process of performing the task or the resulting product is observed and scored. "Performance assessment" is a broad term which "covers many different types of testing methods that require students to demonstrate their competencies or knowledge by creating an answer or product" (Office of Technology Assessment, p. 17). Common forms of performance assessment include doing mathematical computations, writing an extended essay, conducting an experiment, developing a portfolio, conducting an exhibition, or presenting an oral argument. Performance assessment is often used interchangeably with the term *authentic assessment.* However, according to Meyer (1992), the two terms are not the same and need to be differentiated. "In an authentic assessment, the student not only completes or demonstrates the desired behavior, but does it in a real-life context. 'Real-life' may be in terms of the student (e.g., the classroom) or an adult expectation" (p. 93). The phrase *assessment by performance and observation* is preferred because it is more specific and implies methods of assessment that are task oriented or hands on. To distinguish it from the term *performance assessment,* the term *hands-on* performance assessment will be used.

The history of hands-on performance assessment dates back to the industrial training and vocational agriculture schools of the late 1800s (Stockton). In order to prepare youth for the world of work, these schools adopted the project method of teaching. Here, students were tested not by pencil and paper, but by their ability to create a product or demonstrate what they have learned. The rise in public education, the need to Americanize new immigrants into American society, and the development of psychometrics in early 1900s overshadowed hands-on performance assessment and gave birth to the mental testing movement. However, hands-on performance assessment has once again surfaced.

As of 1992, there were seven states that established performance assessment programs. These include Arizona, California, Connecticut, Kentucky, Maryland, New York, and Vermont (Office of Technology Assessment). The reemergence of this form of assessment can be attributed to advances in cognitive research, the increase in the number of constructivist or project-based science curricula funded by the National Science Foundation and other donors, as well as the need for better ways of determining student achievement and school accountability (Shavelson, Baxter, & Pine). The use of hands-on performance-based assessment as a viable option for assessing students' science process skills has been suggested by many scholars (Greig, Wise, & Lomask; California Department of Education; Doran & Hejaily, 1992; Germann, 1992; Tetenbaum; Pizzini, 1992; Shavelson, Baxter, & Pine; Hein, 1987; Educational Testing Service; Black).

In developing hands-on performance assessment, three factors need to be considered: (1) definition of performance tasks desired, (2) preparation of performance tasks, and (3) development of a scoring system (Stiggins, 1994).

The definition of performance tasks desired should be the first consideration in the design of any performance assessment program. This means that the assessor must be very clear about what the desired performance task is and be able to communicate it to the student. Some tasks may require that the student perform an act or behave in a certain way under observation, while other tasks may require the student to produce a product, which is then judged according to a certain set of criteria. One way of making sure that the task is well defined is to conduct a task analysis. A task analysis is a process of breaking down a task into smaller components and putting it together again (Mager & Beach, 1967). By conducting a task analysis, one may be able to isolate those tasks that are important and critical for the accomplishment of the task from those that are insignificant or routine. In addition, it tells the assessor where the difficulties in carrying out the task are, and whether he or she needs to have a single task or a series of tasks to be included in the performance assessment. Once the task has been defined, the assessor then needs to specify what the criteria are for the evaluation of the performance task. Stating the performance criteria clearly allows both students and teachers to share a common understanding about expected outcomes. It acts as a contract on which to make reference.

There is no one way of constructing performance tasks. The procedure to be used will depend on purpose, context, subject matter, and resources. Performance tasks may be constructed so that they provide the student with a statement of the task, list of materials, tools, and equipment required, illustrations, detailed step-by-step instructions, and questions (see figure 6.4).

This type of structured performance assessment task is often designed to elicit a product response. The student taking the assessment has limited options as to how to perform the task. All he or she has to do is to follow directions. Another approach to performance task construction is to make it open-ended as shown in figure 6.5. Open-ended performance task allows the student to develop a plan or strategy for tackling the task. He or she may choose any sequence of steps to arrive at a solution.

Some researchers, such as Shavelson, Baxter, and Pine, have offered five guidelines for constructing performance assessment tasks. They are summarized as follows:

1. Hands-on performance assessment in science should capture student's scientific understanding, critical reflection, problem solving, and provide opportunity for creativity.
2. It should incorporate the use of manipulative and experimental apparatus.
3. It should take advantage of advances in computer technology.

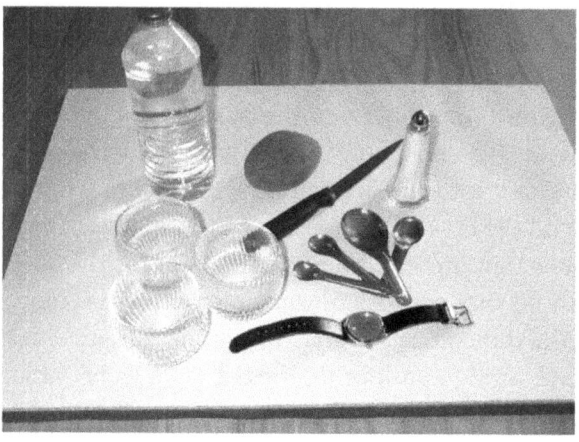

MATERIALS

You have a piece of potato, a bottle of water, three glasses (cups), sodium chloride (Na+Cl-), a set of measuring spoons, a knife, and a watch.

INSTRUCTION

You have 60 minutes to complete this task. Perform an experiment and answer the following questions:

ASSESSMENT TASK/QUESTIONS

1. Find out what happen to the potato when the Na+Cl- concentration is at zero.

2. Find out what happen to the potato when the Na+Cl- concentration is very low.

3. Find out what happen to the potato when the Na+Cl- concentration is very high.

4. Write your results on the worksheet provided. Leave your worksheet on your desk.

Source: Colley, 2008, originally published in *The Science Teacher*, November 2008. (Copyright © National Science Teachers Association, all rights reserved.)

Figure 6.4. Structured Performance Task.

4. It should take advantage of advances in cognitive research.
5. It needs to be linked up with curriculum reform.

There are two ways of scoring hands-on performance tasks, namely by using a holistic scoring criteria (as described earlier under "Essay") or by using a scoring protocol or form, such as the one developed by Baxter, Shavelson, Goldman, and Pine. Holistic scoring procedures vary depending on the focus

> **THE BIAGAM DILEMMA**
>
> **CASE:** The country of Biagam's only natural resource is a fertile land and a beautiful river. The country imports most of its food, even though it can grow all the food it needs. The government of Biagam, with help from the Global Bank, wants to build three dams on the Biagam River to supply water for irrigation and hydroelectric power. This will allow the country's farmers to grow food throughout the year, thus eliminate hunger and dependence on food imports. The hydroelectric power will provide electricity for manufacturing industries, which will create jobs and stimulate economic growth. Although many people in Biagam welcome the idea, some citizens and environmentalists fear that it will lead to an increase in fly-transmitted diseases, such as malaria and trypanosomiasis, destruction of aquatic plant species, fisheries habitat, and flooding of town and cities near the river.
>
> **ASSESSMENT TASK/QUESTION:** You are a scientist called upon to study the Biagam River Watershed to help the government and people decide whether to go ahead with the Biagam River Dam. Explain all the steps you would take to carry out your investigation.
>
> Source: Colley, 2008, originally published in *The Science Teacher*, November 2008. (Copyright © National Science Teachers Association, all rights reserved.)

Figure 6.5. Open-ended Performance Task.

of the performance task and the type of rating scale used. For instance, the California Golden State Examination holistic scoring for laboratory performance tasks focuses on mastery of scientific methods, problem solving, and understanding of scientific concepts. Student tasks are rated on a scale of 0 to 6 (California Department of Education).

The Connecticut Academic Performance Test focuses on science process skills, such as the ability to define a problem, design an experiment, reason logically, make inferences and conclusions that are consistent with observation, perform with minimum errors or misconceptions, and communicate clearly. Students are rated on a qualitative scale such as: Excellent Performance, Proficient Performance, Marginal Performance, and Unsatisfactory Performance (Greig, Wise, & Lomask). In the Alberter Performance-Based Assessment for Science, the holistic scoring criteria focus on problem solving or inquiry and communication. The rating scale used for judging students work is both quantitative and qualitative. It is presented as follows: 3 = Beyond, 2 = At Grade, 1 = Not yet at Grade, and 0 (Hall, 1993). Tamir (1974) suggests the use of a predetermined key and the assigning of relative weights to each skill (in percentage points) such as manipulation = 10, self-reliance = 10, observation = 15, experimental design = 25, reporting and communication = 15, and reasoning and application = 30.

The use of scoring protocol or forms requires the service of trained observers. The training of observers to score performance tasks has been discussed in detail by Baxter, Shavelson, Goldman, and Pine. In their study of a procedure-based scoring for hands-on science assessment, the researchers note, "In scoring hands-on performance, we found that a score could not be assigned without knowing the particular method or approach to experimentation used by the student" (p. 2). This means that observers must not only have a background in the subject matter, but need to try out the performance task as well. In addition, observers should be provided with the opportunity to observe and score multiple samples of performance tasks and reflect on the process. A reliability check could be developed by randomly selecting the scored performance tasks and rescoring them to measure consistency (Hall, 1993).

Strengths of Assessment by Performance

Technical. The design of hands-on performance assessment allows teachers to ask critical questions about curriculum, instructional goals, and teaching. By asking critical questions about such issues, teachers do a better job in relating what is taught to student outcomes. Although the scoring of hands-on performance assessment requires the use of outside judges, studies have shown that hands-on performance assessment can be judged by a single rater with a high degree of reliability (Shavelson & Baxter, 1992; Shavelson, Baxter, & Pine).

Curricula. Hands-on performance assessment allows the student or test taker to construct his or her own answers as opposed to choosing from a group of answers. Both student and teacher are made aware of the skills and knowledge to be learned as well as the criteria for judging performance. Furthermore, hands-on performance assessment relies on multiple forms of testing. Students are judged not at one time, but over time. The work of students is compared to their performance assessment and not to other students' performance.

Cognitive research (Gardner, 1985) indicates that most learning goes on within an active, rather than a passive, context and "that children construct knowledge from their actions on the environment" (Wadsworth, 1989, p. 156). Hands-on performance assessment is, therefore, suitable for assessing nearly all types of learning because it allows students to demonstrate their competency in ways compatible with their learning experience. This method of assessment is particularly applicable for assessing science process skills such as problem solving, design of experiments, analysis and interpretation of data, critical thinking, use of instruments, and drawing conclusions. Hands-on performance assessment can provide teachers or assessors with information on what students have or have not been taught (Shavelson & Baxter).

Social. A study of hands-on performance assessment suggested that one of its social strengths is that it fosters strong commitment and passion among teachers (Aschbacher). The way hands-on performance assessment fosters commitment and passion among teachers can only be understood when placed in context. For instance, Aschbacher tells a story about a teacher who had been practicing hands-on performance assessment for several years, while most of her colleagues used traditional methods of testing. Although she tried to convince her colleagues to adapt her approach, she did not succeed. However, she was sustained because of her own commitment and passion, as well as the commitment of other teachers from different schools in similar situations.

Another social strength of hands-on performance assessment is that it offers students choices in the selection of

> topics or problems, to create and execute unique procedures, or to sequence aspects of performance electively. There is a belief that this element of choice will motivate students to perform well and may account for cultural differences in a more equitable manner (Gordon, 1991) than occurs in standard testing settings. (Baker, O'Neil, Jr., & Linn, 1993, p. 1212)

Economic. Hands-on performance assessment may be economically feasible in a situation where a school is interested in piloting this type of assessment on a small scale, using grant money.

Weaknesses of Assessment by Performance

Technical. A major technical weakness of this method of assessment is that it is labor intensive. In order for a hands-on performance assessment task to be developed, both mental and physical labor needs to be invested. Mental labor investment involves the thinking and documenting of the performance task, while the physical labor investment involves the gathering of materials and resources, constructing, and field testing of the performance task. All these listed activities take time.

Another challenge facing hands-on performance assessment is the standardization of scoring procedures. Evidence from research suggests that there is wide variability in the development of scoring procedures for hands-on performance assessment (Baker, O'Neil, Jr., & Linn; Office of Technology Assessment). Performance tasks vary greatly in terms of the subject matter they address, and the theories and procedures they invoke. Due to this variability, it is difficult to estimate student performance on a particular task and setting when the tasks are limited. According to Shavelson and Baxter, to obtain a realistic picture of a student's science performance, the student must perform numerous investigations, approximately between 10 and 20. Because large-scale performance assessment is costly in terms of time and money, task sampling poses a major problem.

Social. Hands-on performance assessment has great potential not only as a tool for assessment reform but also for education reform. However, some researchers (Baker, O'Neil, Jr., & Linn) claim that a better research base is needed to judge the merits and demerits of hands-on performance assessment. They note that the expectation for this assessment method is very high and, as experience has shown with earlier educational reforms, the potential benefits may be lost when expectations are too high and unrealistic.

Implementing hands-on performance assessment at the district or state level requires some amount of conscientization among teachers and parents. This is because the majority of these populations are used to Standardized Multiple-Choice, Paper and Pencil (SMPP) assessment- and similar ways of measuring students' abilities. In order for it to work, teachers and parents must buy into the idea. An "awareness campaign" will often have to be launched to inform those who will be affected about the use of, and the benefits and limitations of, hands-on performance assessment. Without these efforts, success in implementing hands-on performance assessment is dubious.

Economic. There is some hesitation on the part of the education community to use hands-on performance assessment on a large scale because it takes more time to administer and score. The costs of administering and scoring are 3 to 5 times higher than those of conventional testing methods (Office of Technology Assessment). In the state of California, performance-based tests cost $50.00 per student, compared to $5.00 per student for SMPP (Seal, 1993).

Some training of teachers in (1) judging student work, (2) scientific observation, and (3) defining performance criteria is necessary if hands-on performance assessment is to be successful. Training of teachers takes time and costs money.

Chapter Summary

From the above discussion, it is obvious that there are several methods of assessing students' science learning in PBSI environments. These methods were conveniently classified into three groups, namely assessment by presentation or product, assessment by performance and observation, and assessment by choice selection. Each of the methods are designed, administered, and scored using different procedures. Knowing the methods currently used to assess students' science learning in PBSI environments, and their technical, curricula, economic, and social strengths and weaknesses is an essential first step in implementing an effective project-based science learning experience. The most appropriate method for assessing students' science learning in PBSI from the author's point of view are those methods that offer students the opportunity to demonstrate, in real life and time, their competence through performance and the presentation of their work. This point of view is supported by Masters and Mislevy (1993), who note:

> Recent developments in cognitive and educational psychology reveal that most meaningful learning contrasts markedly to the type of learning implied by standard psychometric procedures—those based on item response theory as well as those using classical true-score test theory. The difference is characterized by the discontinuities of real-world learning, as learners reconfigure their knowledge, combine existing skills in new ways, and develop alternative strategies for solving problems. (p. 239)

The main concern that may be raised about the use of performance-based methods is associated with their cost of implementation (i.e., teacher training, class time, administering, scoring, and reporting). These are valid concerns, but they must be seen as challenges and not obstacles.

Food for Thought

The following questions will help the reader reflect on the chapter.

1. Can you differentiate between the following concepts: assessment, testing, and evaluation?
2. What are the main categories of assessment?
3. What are the main methods of assessing students' science learning in PBSI, inquiry-based and problem-based environments discussed in this chapter?
4. Which assessment method would you recommend for assessing students' science learning in PBSI environments?

· 7 ·

PROJECT-BASED SCIENCE INSTRUCTION, AFTERSCHOOL SCIENCE PROGRAMS, AND COMMUNITY ENGAGEMENT

Chapter Overview

PBSI can be implemented in different educational settings. The two most common settings are formal and informal science education. Formal science education refers to science education conducted inside K–16 educational settings governed by local, state, and national educational policies, laws, and regulations. Informal science education settings refer to science education conducted outside of K–16 educational settings that are less restrictive. Examples of informal science education are science education that occurs in the community, afterschool programs, museums, or similar settings outside the classroom. This chapter will focus on the symbiotic relationships among PBSI, afterschool science programs, and the community. The chapter will begin with a description of the purpose and characteristics of afterschool science programs, and how they are connected to PBSI. This will be followed by a discussion of how the local community can be leveraged as a resource and catalyst to promote PBSI and citizen science education. This chapter will end by examining some of the potential benefits and challenges of afterschool PBSI programs and community engagement in such programs.

Purpose of Afterschool Programs

Afterschool science programs serve different purposes. For instance, they could be implemented to provide supplemental instruction in a specific science content area or areas, to provide tutoring in specific science concepts, skills and/or dispositions, to prepare students for college admission, and/or expose students to "real life" scientists and authentic science experiences. Most afterschool science programs serve multiple purposes.

Characteristics of Afterschool Programs

Afterschool science programs share some characteristics in common. They can be summarized as follows:

1. They are goal or purpose driven.
2. They address a specific educational need or challenge.
3. They are mostly designed to serve students who are traditionally underserved by the formal science education system.
4. They are mostly formed through a partnership among school, community, university, and/or professional science organization.
5. They have an open enrollment system usually on a first come, first served basis, needs basis, and/or space availability.
6. They have a low overhead and are usually operated by part-time or volunteer staff, who are motivated by a passion to serve their communities and not necessarily by dollar signs.
7. They rely on different sources of funding.
8. Program sustainability—maintaining service when funding dries out—is a major challenge.
9. They can last for weeks, months, or years.
10. Because of the informal nature of afterschool science programs and funding challenges, they do not always have the human resource capacity or expertise to plan and implement evaluations or studies to determine impact.

There is no recipe for setting up an afterschool PBSI program. The factors to consider and steps to take to implement PBSI described in the *project cycle* in chapters 3 and 4 are also applicable in afterschool science programs. To start an afterschool science program, one needs a clearly defined purpose and an identified population with a felt need. Then sources of funding should be identified to hire part-time staff and purchase materials and services. A venue, space, or location, such as a local park, a school, university laboratories, a community center, farm, or research facilities, should be determined. Sometimes this element could be provided free of charge by partners as an in-kind contribution. An afterschool science program also needs a curriculum guide or program of study that will provide a teaching and learning framework for students, teachers, and partners. There are open-source or free curricula that could be accessed from the Internet or federal government agencies—such as NASA, EPA, FDA, or USDA—that could be adopted or piloted for an afterschool science program. It is helpful to have an advisory board consisting of few committed individuals from the community who are experts and experienced in their field or STEM fields who could offer technical advice and/or financial support to the afterschool science program. There should be an implementation plan with a schedule or calendar of activities, clearly outlined role and responsibilities, and evaluation plan.

Relationships Between PBSI and Afterschool Science Programs

In PBSI environments, students and their teachers carry out projects over extended periods. The planning, implementing, reporting or presenting, and evaluating of these projects takes time. In addition, the logistics of identifying and selecting tools and materials, organizing and managing of students' learning, and scheduling and coordinating group projects is not only time consuming, but it also requires that science teachers demonstrate competencies and expertise in a wide range of pedagogical and science content knowledge and skills. Complicating matters is the fact most curricula in traditional science education settings are highly structured and regulated. Because PBSI is time- and labor-intensive, it thrives well under informal science education settings such as afterschool science programs, where time and labor are not a major issue.

Afterschool science programs like PBSI are goal- or purpose-driven, and because of that they do not operate well under rigid rules; rather the objective is the accomplishment of goals or purpose. It is important to recognize that different goals or purposes will take different amounts of time and labor to accomplish. Afterschool science programs allow students the flexibility and time to pursue projects of interest and relevance to their lives without the constraints of curriculum, state standards, or other factors. Students can learn at their own time and pace and can get one-on-one mentoring or assistance if they are having difficulties. Some students are more likely to be exposed to cutting-edge science practices, tools, technology, and real scientists in afterschool science programs than in traditional science education settings. One of the key elements of PBSI is that learning takes place within a social context and students are required to collaborate as scientists do in the real world. Afterschool science programs share the same elements with PBSI because they take place within a community context, and students have the opportunity to work with experts on projects that may ultimately impact their community.

Community Engagement in Afterschool PBSI

Community in this context refers to the sum total of the members living and/or working in a particular geographical location such as parents or caregivers, elders, youths, teachers, school administrators, university faculty and their students, community activists, local businesses, government agencies, and not-for-profit organizations. The community can be leveraged as a resource and catalyst for promoting PBSI and citizen science education. Private companies, small businesses, and/or local government agencies provide goods and services in communities. For instance, utilities companies provide electricity, water supply, phone, and Internet service; farms and supermarkets provide the food and products we need to exist; local government provides and maintains the physical infrastructure and protects the environment; and insurance companies and local hospitals provide medical and healthcare services to the community. Community members who work in all of the above systems are usually trained and/or experienced in their fields of work. Some are highly trained and regarded as specialist or experts in their field. Although these workers, professionals, and experts are not trained in pedagogy, they have accumulated a lot of scientific knowledge and experience during their careers, which could be put in the service of afterschool PBSI programs.

For example, most communities are served by a local or regional utility company, which is responsible for providing electricity, telecommunications, and related services to their communities. In addition, the utility company employs workers who are often educated or trained in science discipline such as mathematics, physics, electrical engineering, computer science, and/or related disciplines. Some of these workers are also community residents. Suppose a group of community residents, teachers, and university faculty are interested in establishing an afterschool PBSI. Suppose these partners are interested in using the topic of renewable energy as a vehicle for teaching physics and also for meeting the following NGSS standard: "HS-PS3-3: Design, build, and refine a device that works within given constraints to convert one form of energy into another form of energy" (http://www.nextgenscience.org/hsps-e-energy). A collaborative partnership could be formed between the afterschool PBSI program and the utility company or a renewable energy company. From this partnership, we could expect that the utility company/renewable energy company will provide volunteer experts or technical assistance in renewable energy, funding, equipment, and/or internship opportunities to the afterschool PBSI partners, while the afterschool PBSI partners will provide learning space, recruit students and staff, and develop curriculum. The afterschool PBSI and the utility company/renewable energy company would be expected to engage each other in planning, implementing and evaluating, and capacity building to achieve goals and objectives. In addition, it could be expected that in such a partnership, there would be different levels and types of interactions to leverage resources and opportunities that would ultimately benefit both partners. Evans, Abrams, Rock, and Spencer (2001) offer some helpful tips about student-scientists partnerships that are relevant to this discussion. According to the researchers, student-scientist partnerships (SSPs) are

> a type of project-based instruction, wherein students are active participants in a scientific research collaboration between students, teachers and research scientists. SSP provides context-rich, integrated and hands-on approach to teaching the scientific enterprise as well as subject specific content for K–12 students. They also provide the potential for scientific data collection from a broad geographical range that would not be feasible without such partnerships. To optimize the outcomes of both of these two objectives, critical program components must be in place and the trade-offs between educational activities and accurate data collection for research must be clearly delineated. (p. 318)

The message from the above quote is that for partnerships such as the one described above to work, both partners must be engaged and both must mutually benefit. Having a conversation at the beginning of the partnership about expectations and being able to balance needs, opportunities, and trade-offs will promote a healthy and sustainable partnership. Community engagement through partnerships could lead to increased parental involvement and interest in science. Parental involvement and active teachers and students' engagement in afterschool PBSI could lead to students pursuing critical projects supported by their community with the potential to bring about change in the community. Ordinarily, individual citizens' participation and activism in science does not lead to immediate change or action. However, when participation in science spreads to the wider community and is supported by legislators, change is likely to take place. When students, teachers, parents, scientists, experts, and other community members are engaged in projects that are of interest and relevant to their lives, there is a greater chance that their voices will be heard and there will be enough pressure to bring about desired policy changes.

Potential Benefits and Challenges of Afterschool PBSI Programs

The benefits of afterschool science programs have been well documented (Colley & Pitts Jr., 2010; Walker, Wahl, & Rivas, 2005; Frankel, Streitburger, & Goldman, 2005; Fadigan & Hammrich, 2004; Friedman, 2003; Colley & Pitts, 2003; Marsh & Kleitman, 2002; Scott-Little, Hamann, & Jurs, 2002; Eccles & Templeton, 2002; Fusco, 2001). Afterschool science programs benefit all students, particularly students who are usually not well served by the traditional science education system. Students who participate in afterschool science programs have been known to show a wide range of positive outcomes, such as better understanding of science concepts, acquisition of science process skills, and learning science in ways consistent with their own experiences (Colley & Pitts Jr.). In addition, afterschool science participants benefit in the form of improved academic achievement, more engagement in science learning, higher graduation rates, positive dispositions toward science, and greater tendency to pursue careers in science. In a study of African American women scientists and factors that contributed to their academic and professional success, exposure to experientially rich afterschool science programs was the most cited contributing factor. These African American women scientists got hooked on science and decided to pursue science careers after they participated in an afterschool science program or similar out-of-school science experience (Colley & Colley, 2013).

There are potentially three major challenges that teachers, researchers, practitioners, and/or community members interested in implementing afterschool PBSI could face. The first one is reliable sources of funding to purchase materials, food, pay essential support staff, and program services such as transportation, space, and other miscellaneous items. Usually it is not that difficult to find seed monies to start an afterschool PBSI. There are foundations, local businesses, private donors, and faith organizations that provide seed monies to partnerships or organizations to start an afterschool PBSI program. However, most funding is short term and the expectation is that once established, the afterschool PBSI program will be able to sustain itself. This may not always be true and any person or group interested in afterschool PBSI programs should think carefully about this challenge and develop strategies to mitigate it.

The second challenge with afterschool PBSI programs is finding interested and committed partners (teachers, parents, community members, university faculty/students, scientists, etc.) willing to collaborate and start an afterschool PBSI program. Forming a partnership between groups who traditionally have different views and expectations of science education requires hard work, initially to build relationships and trust. Therefore, it is advisable to start with a small group and conduct a pilot first before establishing a full-scale program.

The third challenge has to do with capacity building. Afterschool PBSI, like any other instruction program, requires caring, competent, and experienced science educators to facilitate science learning and mentor students throughout the project cycle. Identifying and recruiting qualified volunteer staff, instructors, scientists, and parents is never easy, and at the initial stages, afterschool PBSI may require some form of strategic partnership with organizations or institutions that have capacity to provide on-the-job training of staff on content and/or pedagogy. Research on afterschool programs in general have shown that although they provide many benefits to participants, they nevertheless suffer from the lack of robust research studies and data to demonstrate program effects (Scott-Little et al.). The "lack of robust research studies and data" problems could be resolved by identifying master's or doctoral candidates in science education looking for possible opportunities to conduct their research. This could be a win-win situation where candidates use a program site to conduct their studies in exchange for sharing their data and findings with their host.

Chapter Summary

This purpose of this chapter was to demonstrate the symbiotic relationships among PBSI, afterschool science programs, and the community. The chapter identified and described the characteristics of afterschool science programs and how they are connected to PBSI. In addition, the chapter includes a description of the process of starting an afterschool science program. The chapter ends with a discussion on how local communities could be leveraged as a resource and catalyst to promote PBSI and citizen science education.

Food for Thought

The following questions will help the reader reflect on the chapter.

1. List one of the main purposes of afterschool science programs.
2. List five characteristics of afterschool science programs.
3. Describe the symbiotic relationship between PBSI, afterschool science, and the community.
4. Describe two challenges to consider before implementing afterschool PBSI.

· 8 ·

RESOURCES FOR PROJECT-BASED SCIENCE INSTRUCTION

Chapter Overview

This chapter will be devoted to resources that teachers and students can use to plan and implement PBSI. The chapter will identify and describe the following resources: library resources, Internet or online resources, museum and informal science education centers, computer hardware and software, scientific tools and technologies, curriculum materials, electronic databases, government agencies, professional science–science education associations, and scientific research centers and organizations. Each resource category will be described with particular reference to when and/or how it is used in PBSI, its availability or how it could be accessed, any special training required, and potential benefits to teachers and their students. Where possible, information on vendor or supplier will be provided.

Using Library Resources

Libraries are always a good place to start when planning for project-based science instruction. The main components of a library include a librarian, a reference section, general collections, journals and periodicals, and online databases. Of these components, the most important is the librarian. Identifying and defining a question is one of the first steps in the project cycle. When students are at this state of the project cycle, it is a good time to visit the library and consult with the librarian. A librarian can help students learn how to conduct searches and can direct them to the appropriate sections that are relevant for the questions they are investigating. In addition, a librarian can also demonstrate how to use the Index of Research Abstract in the different disciplines so that students can familiarize themselves on prior and ongoing research questions.

Another important component of the library is the online databases, which includes journals, periodicals, and books. This aspect of the project is important in that it is helpful for teachers and students to actually spend some time learning how to use the online databases and working as a group to search and identify literature relating to their topics. In addition, they can also learn about citations and referencing the works of others. This is also a good time to learn about the anatomy of a research article and the function of each of these parts.

The American Library Association (ALA), which consists of 11 library associations, including and not limited to the American Association of School Librarians, Association for Library Service to Children, Association of College and Research Libraries, Library and Information Technology Association, and Public Library Association, has developed a set of standards to guide teachers, students, parents, and administrators on the information literacy skills, dispositions, responsibilities, and self-assessment strategies students should know and be able to demonstrate in the twenty-first century. Teachers and students implementing or planning to implement PBSI should familiarize themselves with the relevant standards so that they can maximize the use of their library resources. The ALA website is located at www.ala.org/groups/divs.

Another important library resource for supporting PBSI is the National Science Digital Library (NSDL). According its website, the NSDL "provides high quality online educational resources for teaching and learning, with current emphasis on the sciences, technology, engineering, and mathematics (STEM) disciplines—both formal and informal, institutional and individual, in local, state, national, and international educational settings" (https://nsdl.oercommons.org/nsdl-overview). The NSDL contains resources that cover education levels K–16; about 16 science disciplines; all major education standards, including the national science education standards and the NGSS; and a wide variety of resource types such as lectures, lesson plans, labs, syllabi, assessments, case studies, data, games, articles, references, and books to name a few. For more information and to explore the NSDL, please visit https://nsdl.oercommons.org.

Using Internet or Online Resources

According to Dictionary.com (2015), the term *Internet* refers to "the global communication network that allows almost all computers worldwide to connect and exchange information. Some of the early impetus for such a network came from the U.S. government network Arpanet, starting in the 1960s" (http://dictionary.reference.com/browse/internet). Oxford Dictionaries (2015) defines the Internet as "a global computer network providing a variety of information and communication facilities, consisting of interconnected networks using standardized communication protocols" (http://www.oxforddictionaries.com/us/definition/american_english/internet). Shelly, Cashman, Waggoner, and Waggoner (1997) refer to the Internet as a "world wide group of connected networks (some local, some regional, and some national) that allows public access to information and services" (p. 732). From the above definitions, we could sum up as follows: The Internet is a worldwide social gathering of computer users, connected via computer networks and distributed globally. The Internet works by packaging bits of data (e-mails, files, folders, links, etc.) and sending it from point of origin to point of destination using a communication protocol called Transmission Control Protocol/Internet Protocol (TCP/IP) and the most efficient traffic possible. Teachers and students could be connected to the Internet via their school networks, at their home via a commercial Internet Service Provider (ISP), or via their local public libraries. Although the structure and functions of the Internet are very complex, the main components are browsers (software to navigate the Internet), Hypertext Markup Language (HTML, a markup language used to create web pages and multimedia on the Internet), the World Wide Web (a collection of documents connected by hyperlinks that could be accessed on any computer connected to the Internet), and search engines (software tools used to search for links, documents, web pages, and related materials on the Internet, e.g., Google, Bing, Yahoo).

The beauty of the Internet is that with a click of a button, you can access information from sources that are ordinarily either too expensive or take a very long time to access. However, the Internet can also be a place where one can experience information overload and/or misinformation. The Internet, like most technological inventions, has a negative side. Criminals and ill-intentioned individual, groups, organizations, or states can use it for bad purposes. In order to take advantage of the resources on the Internet, one must have information literacy skills. According to the American Association of School Libraries (2007),

> Information literacy has progressed from the simple definition of using reference resources to find information. Multiple literacies, including digital, visual, textual, and technological, have now joined information literacy as crucial skills for this century. The continuing expansion of information demands that all individuals acquire the thinking skills that will enable them to learn on their own. The amount of information available to our learners necessitates that each individual acquire the skills to select, evaluate, and use information appropriately and effectively. (p. 3)

In using the Internet or online resources, it is important that students are taught to focus on their research questions and avoid being distracted by other information that is not directly related to their question. It is advisable to always have a notebook or logbook to record sites visited, documents identified, and any interesting features of the site. Cautionary note: Although this step in the process is very important, it should not occupy the majority of time.

Museums and Science Education Centers

Museums and science education centers are places that house scientific resources and provide opportunities for conducting scientific activities and explorations. Museums are usually collections of artifacts that address different disciplines in the field of science; while science education centers also hold artifacts, they also provide opportunities for students to engage in hands-on activities. Although there is a difference between the two, their functions can also overlap. National Parks are also included in the definition of science centers. Prior to implementing PBSI, it is helpful for students to visit local museums and/or science education centers where they can get ideas for their projects. Examples of interesting places where students can visit to get ideas include but are not limited to the Liberty Science Center in New Jersey, Museum of Natural History in New York, Yellowstone in Wyoming, Yosemite National Park in California, Everglades National Park in Florida, or the Johnson Space Center in Texas. To make these visits productive, they should be well planned in advance in collaboration with museum or science center personnel.

Computer Hardware and Software

By far one of the most valuable resources in planning and implementing PBSI are computer hardware and software components. In this book the term *computer hardware* is used broadly to mean a computer (commonly knows as a PC or Mac computer) with all it external components that function together as a system (Roblyer, Edwards, & Havriluk, 1997). Computers now come in different shapes and sizes. These include and are not limited to tablets, laptops, and desktops. Recently, mobile devices or smartphones and their apps (applications) are being used to perform similar functions as computers. Computer software, also called computer programs or applications, are a set of programming codes or instructions that allow users to interact and communicate with their computers. Computer hardware and software have both advantages and disadvantages in science education. As already mentioned in chapter 4, microcomputer-based laboratories (a particular computer hardware and software) opened up new opportunities and possibilities for conducting real-time investigations in the science classroom and empower teachers and students to engage in rigorous science learning. In addition, they save time and labor by carrying out routine tasks faster, thus freeing up time for teachers and their students to focus on critical instructional matters. With computers teachers can easily collect and analyze students' assessment data to inform their science teaching. Computer hardware and software can serve as assistive tools for students who have learning disabilities, thus empowering them and facilitating their inclusion in the science learning process. Despite these advantages, computer hardware and software cost money and not all schools, teachers, and students can afford them. In addition, they could hinder learning by distracting students from their real tasks and/or promoting a false sense of learning that is not anchored in deep understanding of key science concepts. Using computer hardware and software without clear instructional purpose, framework, and understanding of their appropriate applications is like having a malfunctioning car and a fancy mechanic's toolbox, but no instructions on how to use the toolbox and no diagnosis of what is wrong with the car you are about to fix.

To be used effectively in science teaching and learning, teachers and students require basic knowledge and understanding of how to use computer hardware and software, when to use them, and how to maintain and upgrade them in their classrooms. Factors to consider when identifying and selecting computer hardware include amount of memory in gigabytes (GB); the type and speed of processor (fast processors make for better computing); storage capacity; graphics capability to handle images, videos, graphs, and so on; Internet; and wireless capabilities. Other hardware that are useful in PBSI classrooms include and are not limited to smartphones, digital cameras, scanners, printers, SMART boards, data projectors, document cameras, CD players, VHS recorders and players, tape recorders, color monitors, TVs, slide projectors, overhead projectors, microphones, speakers, storage media and devices, digital calculators, and probeware.

The amount and type of software required in conducting PBSI will vary depending on the type of hardware available and projects students are pursuing. However, in general, the software that are commonly used in PBSI environments can be categorized as follows.

1. *Systems software.* These are software that operate computers, tablets, and mobile devices. They are usually preinstalled and come with the new hardware. Examples of systems application software are Macintosh operating system (Apple), Windows operating system (Microsoft), Linux (Linux Foundation), and Android (Google). Most systems software can be automatically upgraded via online protocol or manually via software update features on most computer hardware. However, computers, tablets, and mobile devices that are not outfitted with the latest systems software could be challenging to use in PBSI because of compatibility or functionality issues. Prior to implementing PBSI, it would be wise to check that computers and related hardware have up-to-date systems software or have been fully upgraded.
2. *Application software (or apps).* These are software designed to perform a specific function or functions for the user. Examples include word processing, spreadsheet, presentation (multimedia), database, web browser, antivirus, media player, photo editor, video editor (editing), and videoconferencing, to name a few. Sometimes application software come as a bundle (e.g., Microsoft Office and Apple iWork) or stand alone as in Adobe Acrobat Reader, Apple iMovie, or Windows Media Player.
3. *Learning management software (LMS).* This refers to software that allows educators to plan, develop, delivery, manage, and evaluate learning online and/or face-to-face. There are several LMS, some of which are commercial and some are open-source. Common examples are Blackboard and Moodle. LMS could be used to facilitate

the project cycle in PBSI environments. For instance, some science teachers flip their classes by creating a discussion board, putting all reading materials and assignments online in an LMS, and devote class time to project planning and implementation. Further information on LMS can be found on the following website: www.capterra.com/learning-management-system-software/features-guide.
4. *Database software.* This refers to software used for entering, storing, organizing, managing, retrieving, and generating data reports. Databases are helpful in PBSI particularly during the project implementation phase. With a database software, students can create different data collection and storage templates showing different variables and related fields, where they can enter data directly as they are being collected or upload them as data files. Commonly used database software are Filemaker Pro and Microsoft Access. However, students can create their own databases by using spreadsheet application software such as Microsoft Excel or similar applications. Excel has features that allow users to create rows and columns for data storage. In addition, it has functions such as "sort," "what if," and "basic mathematical rules" that students can use to search and retrieve specific data on their projects.
5. *Statistical software.* Statistical software are used for conducting descriptive and inferential statistical analysis, which includes and is not limited to measuring central tendency and variability, hypothesis testing, design of experiments, determining direction and magnitude of relationships between variables, measuring probability, making predictions, and fitting linear and nonlinear models. Statistics is so central to quantitative thinking and scientific research, yet so poorly understood and implemented in school science in general and PBSI in particular. This is the weakest link in the project cycle chain. Most pre- and in-service teachers and their students can comfortably plan projects, collect data, prepare, and present their findings. However, when it comes to data collection, organization, analysis, and interpretation, they don't always know how to proceed. They rarely are able to make the connection between the use of statistical software as tools for conducting descriptive and inferential statistical analysis and interpretation. Even if they are familiar with statistical software and procedures, they usually lack statistical literacy to interpret their results meaningfully.

This challenge needs to be addressed at the science teacher preparation level, through rigorous coursework in applied statistical thinking and practices, user-friendly statistical software, curriculum materials that focus on data analysis, data modeling, and interpretation, and continual professional development at the in-service science teacher level. Some common statistical software used in schools and colleges and universities that could also be used in PBSI are SAS (Statistical Analysis Systems, developed by the SAS Institute, www.sas.com), JMP (Jump, developed by the business unit of the SAS Institute, www.jmp.com), SPSS (Statistical Package for Social Sciences, now known as IMB SPSS Statistics, www-01.ibm.com/software/analytics/spss/products/statistics), DataDesk (www.datadesk.com), Minitab (www.minitab.com), Statistica (www.statsoft.com), StatPlus (developed by AnalystSoft, a software that interfaces with Microsoft Excel to perform a wide variety of statistical analysis), and Microsoft Excel. It is important to note that although Excel is often categorized as a spreadsheet software, it can perform multiple functions, including statistical, business, and mathematical analysis. In addition, it is readily available in most schools as part of Microsoft Office suite and relatively easy to use. For more information about general statistical software packages used in K–16 educational settings, please see www.amstat.org/careers/statisticalsoftware.cfm.

CHANCE ("Barriers and Opportunities for Statistics in Secondary Education," 2015), one of the magazines co-published by the American Statistical Association (ASA) and Taylor and Francis Group, asked its members what they thought were the greatest barriers and threats to statistics literacy in secondary education and related questions. About 11 statistician or statistics educators responded and their responses were published November 11, 2015, under the title: "What Are the Greatest Barriers and Threats for Statistics in Secondary Education?" Three of the responses are very relevant and instructive to this section and are presented below.

- *Response 1:* "For me, a student only truly experiences statistics and data analysis when they are able to ask and answer questions that interest them using data. To do this, I believe they need to be using a statistical software tool designed for working with data."
- *Response 2:* "The greatest barriers to high-school statistics are students, teachers, and curriculum/assessment writers thinking that computing statistics is statistics. We statisticians know that statistics is the understanding of data in context. Context requires gathering data and analyzing data appropriately."

- *Response 3:* "The two greatest threats are (a) misguided curriculum expectations and materials, and (b) teachers without a strong understanding of statistical big ideas. ... Most secondary mathematics teachers have minimal preparation for teaching statistics, with few opportunities to learn about how to support statistical learning. When their own learning has been dominated by procedural and theoretical exercises at the expense of experiencing the wonder of asking important questions and engaging in a statistical process, teachers lack the foundation and rely on their frequently lackluster textbooks, typically devoid of opportunities for statistical inquiry."

Although the above responses are not representative of the population of statisticians and statistics educators and could be interpreted as gross generalizations, it is clear from the three responses that the perceived challenges to developing statistics literacy that promote deeper understanding and applications of statistics are lack of experience with appropriate statistical software, poor teacher preparation on the subject, and overemphasis on mathematical computation. To conduct PBSI successfully, teachers and their students should develop knowledge and understanding of basic statistical theories and their applications in the STEM fields. In addition, they must demonstrate competencies in using statistical software to analyze and interpret project data. Kwasny (2015) recommends the following resources for teachers, students, parents, administrators, and counselors: the ASA, the Mathematical Association of America (MAA), National Council of Teachers of Mathematics (NCTM), Consortium for the Advancement of Undergraduate Statistics Education (CAUSE), the National Institute of Statistical Sciences (NISS) and the federally funded Project-SET: Statistics Education for Teachers.

6. *Modeling software.* Refers to software used for representation, prediction, and simulation. EduTech Wiki (2015), a website dedicated to educational technology (founded by Dr. Daniel K. Schneider, an associate professor in the Faculty of Psychology and Education, University of Geneva) identified three types of modeling software: microworlds (allows teachers and students to explore different phenomena through computer programming, e.g., LOGO programming environments; see Papert, 1980 for more information), mathematical system simulations, special simulation software (e.g., climate and molecular modeling), and 3-D modeling software (e.g., CAD, computer aided design).

Colella, Klopfer, and Resnick (2001), on the other hand, provide an alternative, yet similar, classifications of models into illustrative, analytical, and simulations. Illustrative models, according to the authors, are models that provide the learner with a visualization of a particular scientific system or phenomena. Analytical models take the form of $y = a+b(c)$ and are mathematical representations whose purpose is to "generate solutions that predict behaviors of systems based on a given set of conditions" (Colella, Klopfer & Resnick, p. 10). Unlike analytical models, the objective of simulation models is to describe "the underlying mechanisms and let them run over time to see what happens. With simulation models, it is easier to incorporate random and probabilistic events, reflecting important aspects of the world around us" (Colella, Klopfer, & Resnick, p. 10).

Modeling is one of the cross-cutting concepts mentioned in the NGSS, and in PBSI students can engage in modeling to explain their project data, make cause-effect inferences, or predictions about the variables they are investigating. There are different modeling software available from private vendors and open sources that could be used in PBSI. Some examples are StarLogo Nova, STELLA, and CODAP (Concord Consortium, http://concord.org). StarLogo Nova is the latest version of StarLogo, which was originally developed by Mitchel Resnick, a professor at the MIT Media Laboratory, to introduce teachers and students to programming and model building in science and mathematics. According to the developers, the StarLogo Nova software is a "programming environment that lets students and teachers create 3D games and simulations for understanding complex systems" (http://education.mit.edu/portfolio_page/starlogonova/). With the StarLogo Nova software, users can create, run, and share models using their browser. StarLogo Nova is free for teachers and students and can be accessed at http://www.slnova.org/. STELLA is a modeling software that "allow you to communicate how a system works—what goes into the system, how those inputs impact the system, and what are the outcomes" (http://www.iseesystems.com/softwares/STELLA-iThink.aspx#). STELLA was developed by isee Systems, formerly High Performance Systems Inc. (http://www.iseesystems.com/). Unlike StarLogo Nova, STELLA is a commercial product and a trial version can be downloaded from the company's website. There are two types of STELLA: a professional version for scientists, engineers, and researchers, and a nonprofessional version for K–12 educators. Site and individual licenses for educators are available. CODAP is an interactive, web-based data analysis software developed by the Concord Consortium (http://concord.org) to support student-centered, inquiry-based, or PBSI learning environments. The software could be used to perform basic exploratory data analysis, and because it is web-based, teachers and students engaged in PBSI can collaborate in the data analysis and interpretation process. CODAP is open source and available for free at http://concord.org/projects/codap.

Closely related to modeling and statistical software are Geographical Information System (GIS) software. According to Zeiler (1999), "The purpose of GIS is to provide a spatial framework to support decisions for the intelligent use of earth's resources and to manage the man-made environment" (p. 2). The USGS (2016) describes GIS as a computer system capable of assembling, storing, manipulating, and displaying geographically referenced information (that is data identified according to their locations). Practitioners also regard the total GIS as including operating personnel and the data that go into the system. (http://webgis.wr.usgs.gov/globalgis/tutorials/what_is_gis.htm.). In this book, a simplified definition will be used as follows: GIS is a collection of different maps (locations, satellite, aerial, road, contour, physical features, climate, resources, etc.) and data tables with a wide variety of variables that are linked to the maps. GIS has applications in many fields, including STEM, social, natural, and life sciences. GIS software allows teachers and students in PBSI to ask questions about relationships, formulate and test hypotheses, model and solve problems, visualize and analyze data, and generate reports and maps. With GIS software, teachers and students can also engage in interdisciplinary inquiry and collaborate across time and space. One of the most popular GIS software used in schools and colleges today in the United States is ArcView GIS (now known as ArcGIS). ArcGIS was created by Environmental Systems Research Institute (ESRI, www.esri.com) for nonprofessional users using desktop applications. In the early 1990 ESRI started an outreach to K–16 and through this initiative, a variety of programs, curriculum materials, and applications in education have been developed. A good resource for teachers and students who want to incorporate ArcGIS in their PBSI projects is GeoNet, a GIS education community created by ESRI and located at https://geonet.esri.com/community/education/overview.

Although the modeling software described above can serve as powerful tools for PBSI and for achieving some of the requirements of the NGSS, they may not be always accessible to all students and teachers. However, Microsoft Excel has been used by mathematicians, engineers, and scientists to model different situations and problems. Those interested in using Excel as a modeling software might want to review the following book: *Computerized Environmental Modeling: A Practical Introduction Using Excel*, by Hardisty, Taylor, and Metcalfe (1993).

Scientific Tools and Technology

Forcier (1999) defines a tool as "a medium for completing some work and extending the user's ability" (p. 34). The term *scientific tools* is used here to mean devices, equipment, or aids that scientists use in pursuit of scientific questions or objectives. Technology means different things to different people. To some, the term *technology* refers to the cultural artifacts invented by a particular civilization. This could include, but is not limited to, farming, architectural, or transportation practices, and so forth. To others, technology is simply the gadgets, devices, and online platforms that we use to communicate and create social media. Yet, technology is also defined as the application of science and engineering principles to produce tangible products, for instance, the application of the laws of physics and chemistry to produce the combustible engine or the laws of genetics to produce drought-resistant plants. All of the above definitions are relevant and apply to the discussion in this section of the chapter.

Scientists and engineers cannot function without tools and technologies. In fact, the history of the development of science and engineering is closely related to the history of the development of tools and technologies in these disciplines. In some cases, the latter has spurred growth, and led to discoveries and the advancement of the particular discipline. Turner (1998) reminds us, "In the 20th century, we have come to accept a vast range of technical, often very complex, equipment for everyday use. Science has become the very substance of our lifestyle. Nevertheless, the appeal of historic scientific instruments remains, and from them much can be learned of the practice and development of science over the centuries" (p. 9).

PBSI cannot be fully implemented without the use of tools and technologies. To benefit from a PBSI experience, teachers and their students must have a direct experience with scientific tools and technologies. When students use scientific tools and technologies, they encounter a very intimate experience with science. They see how science works. They see the challenges and the promises that science holds. In addition, using tools and technologies, students are able to understand the underlying principles of science and how they are applied to produce tangible results. By having a hands-on experience with particular tools and technologies and seeing how they work, students are

RESOURCES FOR PROJECT-BASED SCIENCE INSTRUCTION   165

more likely to become motivated to pursue science careers. When students use tools and technologies in a PBSI environment, they not only engage their minds, they also engage their psychomotor skills (hands-on, minds-on). It is impossible to list all the tools and technologies that students and teachers will need in a PBSI environment. However, what follows are examples of tools and technologies that have been used in PBSI and reported in K–16 science education publications. The information on vendors and products is included only for convenience and does not represent an endorsement by the author.

Microcomputer-based labs (MBL) (see chapter 4)

- PASCO Scientific (www.pasco.com)
- Vernier Software and Technology Company (www.vernier.com)

Environmental testing kits (soil, water, and air quality)

- PASCO Scientific (www.pasco.com)
- Vernier Software and Technology Company (www.vernier.com)
- LaMotte (www.lamotte.com/en/education)
- Scientifics Direct (www.scientificsonline.com)
- Sensidyne, a supplier of air sampling tools (www.sensidyne.com)
- Gastec Corporation, a supplier of gas detector tube system and instructional materials (www.gastec.co.jp/english)
- EPA science and technology website (www.epa.gov/research/methods-models-tools-and-databases)

Mobile weather station

- Vernier Software and Technology Company (www.vernier.com/products/sensors/weather-stations)
- Ambient Weather (www.ambientweather.com)
- WeatherHawk (www.weatherhawk.com/education)

DNA extraction kit

- Edvotek (www.edvotek.com)
- Carolina (www.carolina.com)
- Fisher Science Education (www.fishersci.com/us/en/education-products.html)
- Flinn Scientific Inc. (www.flinnsci.com)
- Educational Innovations, Inc. (www.teachersource.com/category/biology-dna)

Mobile devices
  Tablets:

- http://einsteinworld.com
- https://tabletsinscienceeducation.wordpress.com
- http://www.intel.com/content/www/us/en/education/right-device/tablets-for-education.html?cid=sem43700009484896538&gclid=CLS1i-9yBncsCFZNZhgodz3IPBg&gclsrc=aw.ds
- www.edupad.com
- http://www.igi-global.com/book/tablets-education-integrated-experiences-implications/104645 (*Tablets in K–12 Education: Integrated Experiences and Implications*, by An, Alon, and Fuentes, 2014)

Smartphones:

- https://learningcenter.nsta.org/product_detail.aspx?id=10.2505/4/tst15_082_09_32 (*Turning Your Smartphone into a Science Laboratory: Five Challenges That Use Mobile Devices to Collect and Analyze Data in Physics*, by Vieyra, Vieyra, Jeanjacquot, Marti, and Monteiro, 2015)
- http://www.science-on-stage.de/download_unterrichtsmaterial/iStage_2_Smartphones_in_Science_Teaching.pdf (*iStage 2: Smartphones in science teaching*)

Calculators:

- Edutopia (www.edutopia.org/technology-integration)
- PASCO Scientific (www.pasco.com)
- Vernier Software and Technology Company (www.vernier.com)

High-altitude ballooning tool kit:

- High Altitude Science (www.highaltitudescience.com)
- HobbySpace (www.hobbyspace.com/index.html)
- Amateur Radio High Altitude Ballooning (www.arhab.org/index.php)
- National Space Grant Foundation (www.spacegrant.org)

Digital microscopes:

- AmScope (www.amscope.com)
- The Microscope Store (www.microscope.com)
- Carolina (www.carolina.com)

- Fisher Science Education (www.fishersci.com/us/en/education-products.html)
- Leica Microsystems (www.leica-microsystems.com/applications/education)
- Flinn Scientific Inc. (www.flinnsci.com)

Multimedia tools (video, audio, still images, sound and graphics)

- There are several options for integrating multimedia into PBSI. You can purchase a standalone video, audio, photo, or sound recording device or a combo. Most smartphones today are equipped with multiple multimedia recording devices and this makes them appropriate and easy to use in PBSI. However, some smartphones require apps and/or other peripherals to fully download and utilize their multimedia files (video, photos, music, etc.). Microsoft PowerPoint is a common multimedia software that students use to develop and present their project results and findings in PBSI settings. Because Microsoft PowerPoint comes with Microsoft Office, most schools in the United States have a site license, and students can share files or upload it on a network for viewing. The Journalist Education Association (JEA) website has information, including a compilation of multimedia tools, that is available to students and educators who are interested in creating their own media. The website is located at www.jeadigitalmedia.org/multimedia-tools.

Robotics:

- The Robotics Alliance Project is a NASA-supported website and a clearinghouse of robotics-related educational materials for K–12 science education (http://robotics.nasa.gov/). The site contains information on summer institutes, grant information, robotic challenges, links, and activities for students.
- LEGO Education (part of the LEGO Group, one of the world's largest manufacturers of play materials) is a for-profit company that provides robotics tools and materials, professional development for STEM educators, maintains a blog and online community, and helps educators with grant writing. The company's website is https://education.lego.com/en-us/lesi/.
- Southwest Robotics in Science Education (SWRiSE, www.southwestrobotics.org/) is a nonprofit organization whose purpose is to "provide avenues for students, teachers, and communities as a whole

to incorporate science and especially robotics into their lives. We are particularly interested in providing opportunities to under-served and low-income communities where students have little opportunities to learn about robotics" (http://www.southwestrobotics.org/who-we-are).

Telescopes:

- http://hubblesite.org/ (Hubble website)
- http://amazingspace.org/ (the Space Telescope Education Program, an education group of the Space Telescope Science Institute's Office of Public Outreach. A clearinghouse for education resources based on the Hubble and James Webb space telescopes)
- www.telescope.com/home.jsp (Orion telescope and binocular site)
- http://galileoscope.org/ (Galileoscope, a low-cost telescope kit developed through collaboration of science educators, astronomers, and engineers to promote STEM education)
- www.toysrus.com/family/index.jsp?categoryId=68162616&sr=1&origkw=telescopes (Toys R Us website for telescopes)

Curriculum and Instructional Materials

One of the misconceptions about PBSI is that students spend most of their time engaged in process, and not content. In addition, students taking responsibility for their own learning sometimes conjures the image of students learning without adult supervision where anything goes. On the contrary, PBSI requires intensive instructional planning, curriculum development; use of tools, materials, and technologies; development of appropriate performance-based assessments; scoring rubrics; procedures for conducting student presentations; evaluation of students' project experience; and follow-up activities. PBSI cannot be implemented in a vacuum and must be supported by appropriate curriculum and instructional materials that guide teachers and students through the project cycle. These curriculum and instructional materials must be aligned to local, state, and national learning standards and be grade-level appropriate. Because projects have to be relevant to students' lived experience, the teaching and learning materials need to reflect students' interests and learning needs. In addition, the materials must include varying levels of complexity and relevant disciplinary content that will challenge teachers and students as they engage in PBSI.

There are two ways that teachers can engage in the development of PBSI curricula. One way is to do it yourself by using the project cycle as a framework, and for each phase in the cycle, identify and develop or select and adopt from secondary sources relevant materials to support the implementation of each phase. The other way is to join an already existing PBSI curriculum project, collaborate with colleagues to form a PBSI curriculum development team, or form a school-university partnership PBSI curriculum team.

Over the past two decades, the divisions on Research on Learning in Formal and Informal Settings (DRL) and Undergraduate Education, two of the four divisions under the Directorate of Education and Human Resources (HER) of the NSF, have been funding curriculum development projects that promote the teaching and learning of science using a project-based approach in K–16 STEM education. A simple search of the Directorate's awards archive revealed about 29 funded projects focusing in full or in part on project-based science between 1991 and 2013. It is important to note that these are some of the projects focusing on PBSI that the NSF has funded over the years. There could be other funded PBSI–related projects through other divisions or directorates of the agency. Science educators are encouraged to visit the HER Directorate (www.nsf.gov/dir/index.jsp?org=EHR) to learn more about past and current NSF-funded PBSI or related projects. Below is a list of key U.S.-based organizations that conduct research, develop curriculum materials, implement teacher professional development, and promote PBSI in K–16 STEM education.

1. TERC (Technical Education Research Center, www.terc.edu/display/HOME/Home), founded in 1965 in Cambridge, Massachusetts, is a nonprofit educational organization dedicated to promoting science and mathematics education through research, development, and dissemination of innovative instructional materials, implementation of professional development, and use of innovative technologies. Examples of national and international science curricula developed by TERC are the Global Lab Project, National Geographic Kids Network Project, LabNet Project, and Investigations in Number, Data and Space Project, "On Being Explicit"—The Cheche Konnen Center.
2. Concord Consortium (http://concord.org), founded in 1994, is based in Concord, Massachusetts. It is a nonprofit research and development organization whose work focuses on the use of technology to improve mathematics, science, and engineering education. Examples

of curricula projects, tools, and technologies developed by the Concord Consortium can be found at http://concord.org/projects.
3. Edutopia, a website dedicated to what works in K–12 education run by the George Lucas Educational Foundation. One of the areas of focus for Edutopia is project-based learning. See www.edutopia.org/project-based-learning for more information.
4. Buck Institute for Education (BIE) is a nonprofit organization whose purpose is to promote the use of project-based learning through teacher professional development and the development and dissemination of innovative PBSI curriculum materials. See http://bie.org for more information.
5. The GLOBE Program (Global Learning and Observations to Benefit the Environment, www.globe.gov) is an international science and education program with the following mission: "To promote the teaching and learning of science, enhance environmental literacy and stewardship, and promote scientific discovery" (http://www.globe.gov/about/overview). The GLOBE Program was founded by the U.S. government in 1994 on Earth Day. Former Vice President Al Gore was very instrumental in its development. The main idea behind the project is to create a worldwide community of students, teachers, scientists, and citizens collaborating to monitor their local, national, and global environments. The outcome of the project is to promote global scientific literacy and environmental stewardship. The GLOBE Program is funded by multiple U.S. government agencies including the following: NASA, NSF, National Oceanic and Atmospheric Administration (NOAA), and Department of State. Teachers, educational administrators, or scientists can join the GLOBE Program; information on how to become a member is available at http://www.globe.gov/join/benefits-of-joining. Teachers and students from international countries can also participate in GLOBE through bilateral agreements between the U.S. government and governments of partner nations. Inquiries can be sent via online form to http://www.globe.gov/support/contact.
6. It's About Time (IAT) (www.iat.com) is a for-profit organization that researches, develops, and markets project-based K–16 STEM curricula, as well as provides professional development for STEM teachers. Specifically, IAT provides courses, curricula, technologies, and videos to support PBSI. IAT project-based curricula were developed originally

with support from the NSF. For more about IAT PBSI curricula, visit www.iat.com/courses.

Scientific Electronic Databases

Scientific electronic databases are digital warehouses for large data sets on a wide range of scientific topics. Most of this data comes from publicly funded research, are in the public domain, and can be accessed online through appropriate procedures. Some international scientific organizations also collect various types of scientific data that can be accessed by scientists and educators around the world for education and/or research purposes. The United Nations (UN) and its agencies collect data on a global scale on different scientific subjects, themes, issues, indicators, and variables. In addition, the UN produces special and annual reports that contain valuable data and are made available to the global public. Other than the UN, there are also regional and international bodies that represent different disciplines of science. These bodies promote, support, and fund scientific research and development around the world. They also maintain databases on scientific topics such as energy, climate change, global health, STEM education, food, agriculture, oceanography, space, and earth sciences. Electronic databases are an underutilized resource in PBSI instruction. With the right type of computers, software, and access to the Internet, students could pose scientific questions and use these databases to answer their questions. The experiences and findings from this type of science research can have implications far beyond students' own learning context. Some interesting scientific topics and websites that students and teachers engaged in PBSI can explore to mine scientific data are as follows (this is a partial list).

United Nations: http://www.un.org
United Nations International Children's Fund (UNICEF): www.unicef.org
United Nations Educational Scientific & Cultural Organization (UNESCO): http://www.unesco.org
UN Environmental Program: http://www.unep.org
UN World Health Organization (WHO): http://www.who.org
UN Food & Agriculture Organization: http://www.fao.org
UN World Meteorological Organization: http://www.wmo.int
Intergovernmental Panel on Climate Change: http://www.ipcc.ch
International Council for Science: http://www.icsu.org
Global Network of Science Academies: http://www.interacademies.net

International Union of Biochemistry and Molecular Biology: http://www.iubmb.org
International Union of Microbiological Societies: http://www.iums.org
International Union of Pure and Applied Chemistry: http://www.iupac.org
International Astronomical Union: http://www.iau.org
The Royal Society of the United Kingdom: https://royalsociety.org
The Royal Society of Canada: http://www.rsc.ca
National Research Council: http://www.nationalacademies.org/nrc

Government Agencies

In the United States there are several state and federal government agencies that employ scientists and also fund scientific research. From the research of these scientists are generated a variety of data, some of which are available upon request or are in the public domain. At the federal level, some of the governmental agencies that engage in and/or fund scientific research include but are not limited to the Department of Energy, Environmental Protection Agency, Food and Drug Administration, National Institutes of Health, Department of Agriculture, Department of Health and Human Services, U.S. Geological Survey, NASA, NOAA, and the U.S. Parks Service. Students can visit the websites of these agencies to explore the resources that are made available to K–16 STEM education. Below is a short list of federal government websites that could serve as a resource for PBSI.

U.S. Education of Department: http://www.ed.gov
U.S. Library of Congress: http://lcweb.loc.gov/homepage/lchp.html
National Archives and Records Administration: http://www.nara.gov
U.S. Environmental Protection Agency (EPA): http://www.epa.gov
U.S. Geological Survey: http://www.usgs.gov
U.S. Department of Energy: http://www.doe.gov
National Renewable Energy Laboratory: http://www.nrel.gov
National Institutes of Health: http://www.nih.gov
National Aeronautics and Space Administration (NASA): http://www.nasa.gov
National Science Foundation: http://www.nsf.gov
U.S. Department of Agriculture: http://www.usda.gov
U.S. Food & Drug Administration (FDA): http://www.fda.gov
National Park Service: http://www.nps.gov/index.htm

National Oceanic and Atmospheric Administration (NOAA): http://www.noaa.gov

Professional Science and Science Education Associations

These associations are formed and operated by professional scientists and science educators. Almost every field of science has a specialized professional association whose membership comes from professional scientists, engineers, researchers, and educators. Some of these associations have special divisions devoted to K–16 science education. In addition, they provide services in the form of curriculum resources and professional development for educators to promote literacy in their disciplines. Professional science and science education associations also provide limited grant funding to promote greater participation in their disciplines. The American Association for the Advancement of Science compiled a listing of all its professional science associations affiliates, which includes 252 societies and academies of science; the list can be found at www.aaas.org/aaas-affiliates. Below is a partial list of professional science education associations in the United States.

National Science Teachers Association: http://www.nsta.org
National Council of Teachers of Mathematics: http://www.nctm.org
National Association for Research in Science Teaching: http://www.narst.org
International Society for Technology in Education: http://www.iste.org
Association for Multicultural Science Education: http://amsek16.org
Association for Science Teacher Education: http://theaste.org
Council of State Science Supervisors: http://www.csss-science.org
National Science Education Leadership Association: http://www.nsela.org
Society for College Science Teachers: http://www.scst.org
American Association of Physics Teachers: http://www.aapt.org
National Association of Biology Teachers: http://www.nabt.org
National Association of GeoScience Teachers: http://nagt.org/index.html
National Middle Level Science Teachers Association: http://www.nmlsta.org
American Chemical Society: http://www.acs.org

North American Association for Environmental Education: https://www.naaee.net

American Educational Research Association: http://www.aera.net

Scientific Research Centers and Organizations

Scientific research centers and organizations are places where basic and/or applied research are conducted. They can be privately owned and operated or nonprofit organizations. These research centers and organizations specialize in specific areas of scientific research. For instance, Woods Hole Oceanographic Institution (www.whoi.edu) focuses on oceans; Bell Laboratories (www.bell-labs.com) focuses on information and computing technology; while Cold Spring Harbor Laboratory (www.cshl.edu) focuses on the biological sciences. These centers and organizations are always interested in supporting the teaching and learning of science in K–16 settings. Teachers and students implementing PBSI should identify the research centers located in their communities and tap into this resource, particularly in regards to professional development, PBSI partnerships, guest speakers, and fieldtrips for students.

Chapter Summary

This chapter identified and described resources for planning and implementing PBSI. The main resources for PBSI include, but are not limited to, library resources, Internet and online resources, museum and informal science education centers, computer hardware and software, scientific tools and technologies, curriculum materials, electronic databases, government agencies, professional science and science education associations, and scientific research centers and organizations. Where possible, information on vendors and suppliers and open-source resources are provided and does not indicate endorsement by the author.

Food for Thought

The following questions will help the reader reflect on the chapter.

1. Identify and describe three categories of resources for conducting PBSI.
2. Define the terms *hardware* and *software*.

3. Identify or select two modeling software and explain how you could incorporate them in PBSI.
4. Identify or select two scientific tools or technologies and explain how you could incorporate them in PBSI.
5. What are scientific electronic databases and how would you incorporate them in PBSI?

REFERENCES

Abruscato, J. (1993, February). Early results and tentative implications from the Vermont Portfolio Project. *Phi Delta Kappan, 74,* 474–477.

Adami, G. (2006). A new project-based lab for undergraduate environmental and analytical chemistry. *Journal of Chemical Education, 83*(2), 253–255.

Adams, D. D., & Shrum, J. W. (1990). The effects of microcomputer-based laboratory exercises on the acquisition of line graph construction and interpretation skills by high school biology students. *Journal of Research in Science Teaching, 27*(8), 777–787.

Ahmann, J. S. & Glock, M. D. (1971). *Evaluating pupil growth.* Boston: Allyn & Bacon.

Aikin, W. M. (1942). *The story of the eight-year study.* New York: Harper.

Alozie, N., Eklund, J., Rogat, A., & Krajcik, J. (2010). Genetics in the 21st century: The benefits & challenges of incorporating a project-based genetics unit in biology classrooms. *The American Biology Teacher, 72*(4), 225–230.

Alozie, N., Moje, E. B, & Krajcik, J. S. (2010). An analysis of the supports and constraints for scientific discussion in high school project-based science. *Science Education, 94*(3), 395–427.

American Association for the Advancement of Science. (1989). *Project 2061: Science for all Americans.* New York: Oxford University Press.

American Association of School Libraries. (2007). *Standards for the 21 century learner.* Chicago: American Library Association.

American Psychological Association. (2009). *Publication Manual* (6th ed.). Washington, DC: Author.

An, H., Alon, S., & Fuentes, D. (Eds.). (2014). *Tablets in K–12 Education: Integrated experiences and implications.* Hershey, PA: IGI Global.

Anderson, R. D. (2002). Reforming science teaching: What research says about inquiry. *Journal of Research in Science Teaching, 13*(1), 1–12.

Aschbacher, P. R. (1993). *Issues in innovative assessment for classroom practice: Barriers and facilitators* (CSE Technical Report 359). Los Angeles: University of California, Center for Research on Evaluation, Standards, and Student Testing.

Avraamidou, L. (2013). The use of mobile technologies in project-based science: A case study. *Journal of Computers in Mathematics and Science Teaching, 32*(4), 361–379.

Bagley, W. C. (1921). Dangers and Difficulties of the Project Method and How to Overcome Them: II. Project and purposes in teaching and learning. *Teachers College Record, 22*(4), 288–297.

Baker, E. L., O'Neil, Jr., H. F., & Linn, L. (1993). Policy and validity prospects for performance-based assessment. *American Psychologist, 48*(12), 1210–1218.

Barab, S. A., Hay, K. E., Barnett, M., & Keating, T. (2000). Virtual solar system project: Building understanding through model building. *Journal of Research in Science Teaching, 37*(7), 719–756.

Barak, M., & Dori, Y. J. (2005). Enhancing undergraduate students' chemistry understanding through project-based learning in an IT environment. *Science Education, 89*(1), 117–139.

Barriers and opportunities for statistics in secondary education. (2015). *Chance.* Retrieved from http://chance.amstat.org/2015/11/conversation/

Barron, B. J. S., Schwartz, D. L., Vye, N. J., Moore, A., Petrosino, A., Zech, L., & Bransford, J. D. (1998). Doing with understanding: Lessons from research on problem- and project-based learning. *The Journal of the Learning Sciences, 7*(3–4), 271–311.

Baxter, G. P., Elder, A., & Glaser, R. (1994). *Cognitive analysis of a science performance assessment. Project 2.1 designs for assessing individual and group problem solving. Assessing the validity of existing assessments of problem-solving performance in science: A taxonomy of cognitive processes.* ERIC Number: ED376214.

Baxter, G. P., & Glaser, R. (1998). Investigating the cognitive complexity of science assessments. *Educational Measurement: Issues and Practice, 17*(3), 37–45.

Baxter, G. P., Shavelson, R. J., Goldman, S. R., & Pine, J. (1992). Evaluation of procedure based scoring for hands-on science assessment. *Journal of Educational Measurement, 29*(1), 1–17.

Beaton, A. E., Martin, M. O., Mullis, E. V. S., Gonzalez, E. J., Smith, T. A., & Kelly, D. L. (1996). *Science achievement in the middle school years: IEA's Third International Mathematics and Science Study.* Chestnut Hill, MA: TIMSS International Study Center, Boston College.

Beck-Winchatz, B., & Bramble, J. (2014). High-altitude ballooning student research with yeast and plant seeds. *Gravitational and Space Research, 2*(1), 117–127.

Bell, B., Osborn, R., & Tasker, R. (1992). Appendix A: Finding out what children think. In R. Osborne & P. Freyberg (Eds.), *Learning in science: The implications of children's science.* Auckland, NZ: Heinemann.

Bencze, J. L., & Bowen, G. M. (2007). Student-teachers' dialectically developed motivation for promoting student-led science projects. *International Journal of Science and Mathematics Education, 7*(1), 133–159.

Beneke, S. (2003). Practical strategies. In J. H. Helm & S. Beneke (Eds.), *The power of projects: Meeting contemporary challenges in early childhood classrooms—strategies and solutions*. New York: Teachers College.

Berenfeld, B. (1994). Technology and the new model of science education: The Global Lab experience. *Machine Mediated Learning, 4*(2 & 3), 203–227.

Beveridge, W. I. B. (1957). *The art of scientific investigation*. New York: Vintage.

Bhattacharyya, S., & Bhattacharya, K. (2009). Technology-integrated project-based approach in science education: A qualitative study of in-service teachers' learning experiences. *Electronic Journal of Science Education, 13*(3), 1–26.

Bilgin, I., Karakuyu, Y., & Ay, Y. (2014). The effects of project-based learning on undergraduate students' achievement and self-efficacy beliefs towards science teaching. *Eurasia Journal of Mathematics, Science & Technology Education 11*(3), 469–477.

Black, P. (1987) *Report: National curriculum: Task group on assessment and testing*. London: Department of Education and Science.

Booth, W. C., Colomb, G. G., & Williams, J. M. (1995). *The craft of research*. Chicago: University of Chicago Press.

Bowles, S., & Gintis, H. (1976). *Schooling in capitalist America: Educational reform and the contradictions of economic life*. New York: Basic.

Brandt, R. (1987). On assessment in the arts: A conversation with Howard Gardner. *Educational Leadership, 45*, 30–33.

Brickman, P., Gormally, C., Francom, G., Jardeleza, S. E., Schutte, V. G. W., Jordan, C., & Kanizay, L. (2012). Media-savvy scientific literacy: Developing critical evaluation skills by investigating scientific claims. *The American Biology Teacher, 74*(6), 374–379.

Brown, A. L., & Campione, J. C. (1994). Psychological theory and the design of innovative learning environments: On procedure, principles and systems. In L. Schauble & R. Glasser (Eds.), *Innovations in learning: New environments for education* (pp. 289–325). Mahwah, NJ: Erlbaum.

Brown, A. L., & Campione, J. C. (1996). Guided discovery in a community of learners. In K. McGilly (Ed.), *Classroom lessons: Integrating cognitive theory and classroom practice* (pp. 229–270). Cambridge, MA: MIT Press.

Burns, J. C., Okey, J. R., & Wise, K. C. (1985). Development of an integrated process skills test (TIPS II). *Journal of Research in Science Teaching, 22*(2) 169–177.

California Department of Education. (1993). *Golden State Examination: A guide for teachers and students*. Sacramento, CA: Author.

Carey, S., Evans, R., Honda, M., Jay, E., & Unger, C. (1989). "An experiment is when you try it and see if it works": A study of grade 7 students' understanding of the construction of scientific knowledge [Special Issue]. *International Journal of Science Education, 11*(5), 514–529.

Chang, C., & Tseng, K. (2011). Using a web-based portfolio assessment system to elevate project-based learning performances. *Interactive Learning Environments, 19*(3), 211–230.

Charters, W. W. (1922). Regulating the project. *Journal of Educational Research, 5,* 245–246.

Chin, C., & Chia, L. G. (2008). Problem-based learning tools: Problem-based learning pedagogy and strategies are used to implement project-based science. *The Science Teacher, 75*(8), 44–49.

Chittenden, E. (1991). Authentic assessment, evaluation, and documentation of student performance. In V. Perrone (Ed.), *Expanding student assessment* (pp. 22–31). Alexandria, VA: ASCD.

Cohen, K. C. (Ed.). (1997). *Internet links for science education: Student-science partnerships.* New York: Plenum.

Colaianne, B. (2015). Global warming: Project-based science inspired by the intergovernmental panel on climate change. *The Science Teacher, 82*(1), 37–42.

Colella, V. S., Klopfer, E., & Resnick, M. (2001). *Adventures in modeling: Exploring complex, dynamic systems with StarLogo.* New York: Teachers College Press.

Colley, K. E. (1993). Evaluating the Global Laboratory Project: A model for evaluating project-enhanced science curriculum. A paper presented at the annual conference of the International and Comparative Education Society in Kingston, Jamaica, West Indies, March 13–20.

Colley, K. E. (1997). What's in our water?: A report on the pilot-testing of a telecommunication-based science curriculum unit for elementary grades. The Kids Network: Leveraging learning project. Andover, MA: The Network Inc.

Colley, K. E. (1998). Setting the standard: A model for teaching the New York State learning standards for mathematics, science and technology to prospective teachers. *Impact on Instructional Improvement, 27*(2), 27–34.

Colley, K. E. (2001). Technology-based learning: Using water studies as a basis for an alternative teaching strategy. *The Science Teacher, 68*(6), 49–52.

Colley, K. E. (2005). Project-based science instruction: Teaching science for understanding. *Radical Pedagogy, 7*(2), 1–7.

Colley, K. E. (2006). Understanding ecology content knowledge and acquiring science process skills through project-based science instruction. *Science Activities, 43*(1), 26–33.

Colley, K. (2008). Project-based science: A primer—An introduction and learning cycle for implementing project-based science. *The Science Teacher, 75*(8), 23–28.

Colley, B. M. (2014). Voices from The Gambia: Parents' perspectives on their involvement in their children's education. *Childhood Education, 90*(3), 212–218.

Colley, K. E. & Broderick, M. (1994). Eco-research: A study of student acquisition of science process skills through environmental research and tele-collaboration. Cambridge, MA: TERC, Inc.

Colley, K. E., & Colley, B. M. (2013). *Resilience and success: The professional journeys of African American women scientists.* New York: Peter Lang.

Colley, K. E., & Pitts, W. B. (2003). Afterschool science. *The Science Teacher, 70*(3), 55–59.

Colley, K. E., & Pitts, Jr., W. B. (2010). Project-based after-school science in New York City. In R. Yager (Ed.), *Exemplary for resolving societal challenges* (pp. 19–31). Arlington, VA: NSTA.

Collins, A. (1992a, March). Portfolios: Questions for design. *Science Scope, 15*(6), 25–27.

Collins, A. (1992b). Portfolios for science education: Issues in purpose, structure, and authenticity. *Science Education, 76*(4), 451–463.

Collins, M. A. J. (1984). Improving learning with computerized tests. *The American Biology Teacher, 4*(3), 188–191.

Comer, J. P. (1980). *School power: Implications of an intervention project*. New York: The Free Press.

Cook, K. (2009). A suggested project-based evolution unit for high school: Teaching content through application. *The American Biology Teacher, 71*(2), 95–98.

Cook, K., Buck, G., & Park Rogers, M. (2012). Preparing biology teachers to teach evolution in a project-based approach. *Science Educator, 21*(2), 1–30.

Cook, N. D., & Weaver, G. C. (2015). Teachers' implementation of project-based learning: Lessons from the research goes to school program. *Electronic Journal of Science Education, 19*(6), 1–45.

Crawford, B. A., Krajcik, J. S., & Marx, R. W. (1999). Elements of a community of learners in a middle school science classroom. *Science Education, 8*(6)3, 701–723.

D'Amico, L., Gomez, L. M., & McGee, S. (2015). *A Case Study of Student and Teacher Use of Projects in a Distributed Multimedia Learning Environment*. Proceedings of the Education Testing Service Conference on Natural Language Processing Techniques and Technology in Assessment and Education, Princeton, NJ.

Davis, F. J., Lockwood-Cooke, P., & Hunt, E. M. (2011). Hydrostatic pressure project: Linked-class problem-based learning in engineering. *American Journal of Engineering Education, 2*(1), 43–50.

Delgado-Gaitan, C. (2001). *The power of community: Mobilizing for family and schooling*. Lanham, MD: Rowman & Littlefield.

de los Santos, D. M., Montes, A., Sánchez-Coronilla, A., & Navas, J. (2014). Sol-gel application for consolidating stone: An example of project-based learning in a physical chemistry lab. *Journal of Chemical Education, 91*(9), 1481–1485.

Dewey, J. (1902). *The child and the curriculum*. Chicago: University of Chicago Press.

Dewey, J. (1910). *How we think*. Lexington, MA: D.C. Heath.

Dewey, J. (1916). *Democracy and education*. New York: Free.

Dickinson, G., & Jackson, J. K. (2008). Planning for success: How to design and implement project-based science activities. *The Science Teacher, 75*(8), 29–32.

Dictionary.com (2015). Internet. Retrieved from http://dictionary.reference.com/browse/internet

Dillashaw, F. G., & Okey, J. R. (1980). Test of the integrated science process skills for secondary science students. *Science Education, 64*(5), 601–608.

Doppelt, Y. (2005). Assessment of project-based learning in a MECHATRONICS context. *Journal of Technology Education, 16*(2). Retrieved from http://scholar.lib.vt.edu/ejournals/JTE/v16n2/doppelt

Doran, R. L., & Hejaily, N. (1992, March). Hands-on evaluation: A how to guide. *Science Scope, 15*(6), 9–11.

Draper, A. J. (2004). Integrating project-based service-learning into an advanced environmental chemistry course. *Journal of Chemical Education, 81*(2), 221–224.

Driver, R., Guesne, E., & Tiberghien, A. (1985). Children's ideas in science. Milton Keynes, England: Open University Press.

Driver, R., Asoko, H., Leach, J., Mortimer, E., & Scott, P. (1994). Constructing scientific knowledge in the classroom. *Educational Researcher, 23*(7), 5–12.

Duncan, R. G., & Tseng, K. A. (2010). Designing project-based instruction to foster generative and mechanistic understandings in genetics. *Science Education, 95*(1), 21–56.

Eccles, J. S., & Templeton, J. (2002). Extracurricular and other after-school activities for youth. In W. G. Secada (Ed.), *Review of research in education* (pp. 113–180). Washington, DC: American Educational Research Association.

Educational Testing Service (1987). *Learning by doing. A manual for teaching and assessing higher-order thinking in science and mathematics.* Princeton, NJ: Educational Testing Service, Report No. 17 HOS–80.

Epstein, J. L. (2010). *School, family, and community partnerships: Preparing educators and improving schools* (2nd ed.). New York: Westview.

Erickson, R. C., & Wentling, T. L. (1976). *Measuring student growth: Techniques and procedures for occupational education.* Boston: Allyn & Bacon.

Evans, C. A., Abrams, E. D., Rock, B. N., & Spencer, S. L. (2001). Student/scientist partnerships: A teacher's guide to evaluating the critical components. *The American Biology Teacher, 63*(5), 318–323.

Evans, S. S., Evans, W. H., & Mercer, C. D. (1986). *Assessment for instruction.* Boston: Allyn & Bacon.

Fadigan, K. A., & Hammrich, P. L. (2004). A longitudinal study of the educational and career trajectories of female participants in an informal science education program. *Journal of Research in Science Teaching, 41*(8), 835–860.

Fallik, O., Eylon, B. S., & Rosenfeld, S. (2008). Motivating teachers to enact free-choice project-based learning in science and technology (PBLSAT): Effects of a professional development model. *Journal of Science Teacher Education, 19*(6), 565–591. doi10.1007/s10972-008-9113-8

Finley, F. N. (1986). Evaluating instruction: The complementary use of clinical interviews. *Journal of Research and Science Teaching, 23*(7), 635–650.

Forcier, R. C. (1999). *The computer as an educational tool: Productivity and problem solving* (2nd ed.). Upper Saddle River, NJ: Merrill.

Frankel, S. L., Streitburger, K., & Goldman, G. (2005). *After-school learning: A study of academically focused after-school programs in New Hampshire.* Portsmouth, NH: RMC.

Frazier, D. M., & Paulson, F. L. (1992, May). How portfolios motivate reluctant writers. *Educational Leadership, 49*, 62–65.

Freeman, J. G., Marx, R. W., & Cimellaro, L. (2004). Emerging considerations for professional development institutes for science teachers. *Journal of Science Teacher Education, 15*(2), 111–131.

Friedman, L. (2003). Promoting opportunity after school. *Educational Leadership, 60*(4), 79–82.

Fusco, D. (2001). Creating relevant science through urban planning and gardening. *Journal of Research in Science Teaching, 38*(8), 860–877.

Gallentine, J. L. (1968). Biological drawings: Opinion vs. research. *The American Biology Teacher, 30*, 110–113.

Gardner, H. (1983). *The unschooled mind: How children think, and how schools should teach.* New York: Basic.

Gardner, H. (1985). *Frames of mind.* New York: Basic.

Gardner, H. (1991). *The unschooled mind: How children think and schools should teach.* New York: Basic.

Germann, P. J. (1992, March). A model approach. *Science Scope, 15*(6), 21–22.

Gibbs, G. (2014). Student engagement, the latest buzzword. *Times Higher Education.* Retrieved from https://www.timeshighereducation.com/news/student-engagement-the-latest-buzzword/2012947.article

Giordano, E. (August, 1994). Conversation with Dr. Enrica Giordano, visiting professor at TERC. Institute of Physics, University of Milan, Italy.

Glaser, R., & Silver, E. (1994). Assessment, testing and instruction: Retrospect and prospect. *Review of Research in Education, 20*, 393–419.

Glass, G., McGraw, B., & Smith, M. (1981). *Meta-analysis in social research.* Newbury Park, CA: Sage.

Gordon, E. (Speaker). (1991, November). *Alternatives for measuring performance.* Schools that Work (Video Conference No. 4). Chicago: North Central Regional Educational Laboratory.

Green, K. E. (1991). Measurement theory. In Green, K. E., (Ed.), *Educational testing: Issues and applications.* New York: Garland.

Greene, S. (2013). Mapping low-income African American parents' roles in their children's education in a changing political economy. *Teachers College Record, 115*(10), 1–33.

Greig, J., Wise, N., & Lomask, M. (1994). *The Development of an Assessment of Scientific Experimentation Proficiency for Connecticut's Statewide Testing Program.* Paper presented at the Annual Meeting of the American Educational Research Association in New Orleans, LA.

Gronlund, N. E. (1993). *How to make achievement tests and assessment* (5th ed.). Boston: Allyn & Bacon.

Guesne, E. (1992). Light. In R. Driver, E. Guesne, & A. Tiberghien (Eds.), *Children's ideas in science.* Philadelphia: Open University Press.

Gunstone, R. F., & Mitchell, I. J. (1998). Metacognition and conceptual change. In J. J. Mintzes, J. H. Wandersee, & J. D. Novak (Eds.), *Teaching science for understanding: A human constructivist view* (pp. 133–164). London: Academic.

Gutek, G. L (2001). *Historical and philosophical foundations of education: A biographical introduction* (3rd ed.). Upper Saddle River, NJ: Merrill, Prentice Hall.

Gwynne-Thomas, E. H. (1981). *A concise history of education to 1900 AD.* Lanham, MD: University Press of America.

Hall, G. (1993). *Performance assessment: Science and equity.* Paper presented at the CRESST Assessment Conference, September 13–14. Edmonton, Alberta, Canada: Alberta Education.

Han, S., Capraro, R., & Capraro, M. M. (2015). How science, technology, engineering, & mathematics (STEM) project-based learning (PBL) affects high, middle, and low achievers differently: The impact of student factors on achievement. *International Journal of Science and Mathematics Education, 13*(5), 1089–1113.

Hannes, M. (1921). Project teaching in an advanced fifth grade. *Teachers College Record, 19*(2), 137–148.

Hardisty, J., Taylor, D. M., & Metcalfe, S. E. (1993). *Computerized environmental modeling: A practical introduction using Excel.* Chichester, UK: Wiley & Sons.

Harlen, W. (1984). *Science at age 11. Assessment of performance unit.* London: Her Majesty's Stationery Office.

Harlen, W. (2013). *Assessment and inquiry-based science education: Issues in policy and practice.* Trieste, Italy: Global Network of Science Academies Science Education Programme.

Hein, G. E. (1987). The right test for hands-on learning? *Science and Children, 25*(2), 8–12.

Hein, G. E. (1991). Active assessment for active science. In V. Perrone (Ed.), *Expanding Student Assessment.* Alexandria, VA: ASCD.

Helm, J. H., & Katz, L. (2001). *Young investigators: The project approach in early childhood education.* New York: Teachers College Press.

Hike, N., & Beck-Winchatz, B. (2015). Near-space science: A ballooning project to engage students with space beyond the big screen. *The Science Teacher, 82*(1), 29–36.

Hill, C. & Larsen, E. (1992). *Testing and assessment in secondary education: A critical review of emerging practices.* Berkeley, CA: National Center for Research in Vocational Education (NCRVE), University of California at Berkeley.

Holt, J. (1983). *How children learn.* Cambridge, MA: DaCapo.

Hosic, J. F., & Chase, S. E. (1924). *Brief guide to the project method.* Yonkers-on-Hudson & Chicago: World Book.

Howe, K., & Berv, J. (2000). Constructing constructivism, epistemological and pedagogical. In D. C. Phillips & M. Early (Eds.), *Constructivism in education: Opinions and second opinions on controversial issues.* Ninety-ninth yearbook of the Society for the Study of Education. Chicago: University of Chicago Press.

Hunter, J., Schmidt, F., & Jackson, G. (1982). *Meta-analysis.* Newbury Park, CA: Sage.

Isaac, S., & Michael, W. B. (1979). *Handbook in research and evaluation.* San Diego, CA: Edits.

Jaeger, R. M. (1993). *Statistics, a spectator sport* (2nd ed.). Newbury Park, CA: Sage.

Johnson, J. A., Dupuis, V. L., Musial, D., Hall, G. E., & Gollnick, D. M. (1999). *Introduction to the foundations of American education.* Boston: Allyn & Bacon.

Jones, M. G., & Carter, G. (1998). Small groups and shared constructions. In J. J. Mintzes, J. H. Wandersee, & J. D. Novak (Eds.), *Teaching science for understanding: A humanist constructivist view.* San Diego, CA: Academic.

Juhl, L., Yearsely, K., & Silva, A. J. (1997). Interdisciplinary project-based learning through an environmental water quality study. *Journal of Chemical Education, 74*(12), 1431–1433.

Kanis, I. B. (1991). Lab skills. *The Science Teacher, 58*(1), 29–33.

Kanter, D. E. (2010). Doing the project and learning the content: Designing project-based science curricula for meaningful understanding. *Science Education, 94*(3), 525–551.

Kanter, D. E., & Konstantopoulos, S. (2010). The impact of a project-based science curriculum on minority student achievement, attitudes, and careers: The effects of teacher content and pedagogical content knowledge and inquiry-based practices. *Science Education, 94*(5), 855–887.

Karaçalli, S., & Korur, K. (2014). The effects of project-based learning on students' academic achievement, attitude, and retention of knowledge: The subject of "Electricity in Our Lives." *School Science and Mathematics, 114*(5), 224–235.

Kiefer, A. M., Bucholtz, K. M., Goode, D. R., Hugdahl, J. D., & Trogden, B. G. (2012). Undesired synthetic outcomes during a project-based organic chemistry laboratory experience. *Journal of Chemical Education, 89*(5), 685–686.

Kilpatrick, W. H. (1918). The project method. *Teachers College Record, 19*(4), 319–35.

Kilpatrick, W. H. (1924). How shall we select the subject matter of the elementary school curriculum? *Journal of Educational Method, 4*, 3–10.

Kliebard, H. M. (1986). *The Struggle for the American curriculum 1893–1958*. New York & London: Routledge & Kegan Paul.

Klopfer, L. E. (1973). Evaluation of science achievement and science test development in an international context: The IEA study in science. *Science Education, 57*(3) 387–403.

Knight, P. (1992, May). How I use portfolios of mathematics. *Educational Leadership, 49*, 71–72.

Koretz, D., Stecher, B., & Deibert, E. (1992). *The Vermont portfolio assessment program: Interim report on implementation and impact*. Los Angeles, CA: National Center for Research on Evaluation Standards and Student Testing (CRESST), CSE Technical Report 350.

Koretz, D., Stecher, B., Klein, S., McCaffrey, D., & Deibert, E. (1993). *Can portfolios assess student performance and influence instruction? The 1991–92 Vermont Experience*. Center for Research on Evaluation, Standards, and Student Testing/UCLA Center for the Study of Evaluation Technical Report 371.

Kosinski-Collins, M., & Gordon-Messer, S. (2010). Using scientific purposes to improve student writing & understanding in undergraduate biology project-based laboratories. *The American Biology Teacher, 72*(9), 578–579.

Krajcik, J. (2015). Project-based science: Engaging students in three-dimensional learning. *The Science Teacher, 82*(1), 25–27.

Krajcik, J., Blumenfeld, P. C., Marx, R. W., Bass, K. M., Fredricks, J., & Soloway, E. (1998). Inquiry in project-based science classrooms: Initial attempts by middle school students. *The Journal of the Learning Sciences, 7*(3, 4), 313–350.

Krajcik, J., Czerniak, C., & Berger, C. (1999). *Teaching children science: A project-based approach*. Boston: McGraw-Hill.

Krajcik, J. S., & Layman, J. W. (1992). Microcomputer-based laboratories in the science classroom. In F. Lawrenz, K. Cochran, J. S. Krajcik, & P. Simpson (Eds.), *Research matters—To the science teachers* (pp. 101–108). NARST Monograph, Number 5.

Krajcik, J., McNeill, K., & Reiser, B. J. (2008). Learning-goals-driven design model: Developing curriculum materials that align with national standards and incorporate project-based pedagogy. *Science Education, 92*(1), 1–32.

Kwasny, M. (2015). Statistics in K–12: Educators, students, and us. *Chance*. Retrieved from http://chance.amstat.org/2015/11/statistics-in-k-12/

Labin, S. N., Duffy, J. L. Mayers, D. C., Wandersman, A., & Lesesne, C. A. (2012). A research synthesis of the evaluation capacity building literature. *American Journal of Evaluation*, 33(3), 307–338.

Laffey, J., Tupper, T., Musser, D., & Wedman, J. (1998). A computer-mediated support system for project-based learning. *Educational Technology Research and Development*, 46(1), 73–86.

Lee, J. O., Lee, J., & Lee, E. (2015). Project approach in South Korea: Kimchi/Kimjang. *Childhood Education*, 91(2), 117–122.

Lehman, J. D., George, M., Buchanan, P., & Rush, M. (2006). Preparing teachers to use problem-centered, inquiry-based science: Lessons from a four-year professional development project. *The Interdisciplinary Journal of Problem-based Learning*, 1(1), 76–99.

Lien, A. (1980). *Measurement and evaluation in learning*. Dubuque, IA: William C. Brown.

Lindvall, C. M. (1961). *Testing and evaluation: An introduction*. New York: Harcourt, Brace & World.

Liu, S. (2014). *Implementing project-based learning in physics and statics courses*. Paper presented at American Society for Engineering Education, Paper ID #9331.

Lundeberg, M. A., & Fox, P. W. (1991). Do laboratory findings on test expectancy generalize to classroom outcomes? *Review of Educational Research*, 61(1), 94–106.

Lunetta, V. N., Hofstein, A., & Giddings, G. (1981). Evaluating science laboratory skills. *The Science Teacher*, 48(1), 22–25.

MacTemporary Services, (January, 1993). Personal communication with MacTemporary Services about hiring a Temp. to transcribe interviews. Cambridge, MA.

Madaus, G. F. (1993). A national testing system: Manna from above? An historical/technological perspective. *Educational Assessment*, 1(1), 9–26.

Mager, R. F. & Beach, K. M. (1967). *Developing vocational instruction*. Belmont, CA: Pitman Learning, Inc.

Marsh, H. W., & Kleitman, S. (2002). Extracurricular school activities: The good, the bad, and the nonlinear. *Harvard Educational Review*, 72(4), 464–511.

Marx, R., Blumenfeld, P. C., Krajcik, J., Blunk, M., Crawrord, B., Kelly, B., & Meyer, K. M. (1994). Enacting project-based science: Experiences of four middle grade teachers. *The Elementary School Journal*, 94(5), 517–538.

Marx, R. W., Blumenfeld, P. C., Krajcik, J. S., Fishman, B., Soloway, E., Geier, R., & Tal, R. (2004). Inquiry-based science in the middle grades: Assessment of learning in urban systemic reform. *Journal of Research in Science Teaching*, 40(10), 1063–1080.

Masters, G. N. & Mislevy, R. J. (1993). New views of student learning: Implications for educational measurement. In Frederiksen, N., Mislevy, R. J., & Bejar, I. I. (Eds.), *Test theory for a new generation of tests*. Hillsdale, NJ: Lawrence Erlbaum.

Mattheis, F. E., & Nakayama, G. (1988). *Development of the performance of process skills (POPS) test for middle grades students*. ERIC ED305252.

McIntyer, P. J. (1972). The model identification test: A limited verbal science test. *Science Education*, 56(3): 345–357.

Metcalf, S. J., & Tinker, R. F. (2004). Probeware and handhelds in elementary and middle school science. *Journal of Science Education and Technology*, 13(1), 43–49.

Metz, S. (2015). Project-based science learning. *The Science Teacher*, 82(1), 6.

Meyer, C. (1992, May). What's the difference between authentic and performance assessment? *Educational Leadership, 49,* 39–40.

Miedijensky, S., & Tal, T. (2009). Embedded assessment in project-based science courses for the gifted: Insights to inform teaching all students. *International Journal of Science Education, 31*(18), 2411–2435.

Mills, J. E., & Treagust, D. F. (2003). Engineering education, Is problem-based or project-based learning the answer? *Australasian Journal of Engineering Education, 3,* 1–16. Retrieved from http://www.aaee.com.au/journal/2003/mills_treagust03.pdf

Milner-Bolotin, M., & Svinicki, M. D. (2001). Teaching physics of everyday life: Project-based instruction and collaborative work in undergraduate physics course for nonscience majors. *Journal of the Scholarship of Teaching and Learning, 1*(1), 25–40. Retrieved from http://josotl.indiana.edu/issue/view/135

Mintzes, J. J., Wandersee, J. H., & Novak, J. D. (1998). *Teaching science for understanding: A human constructivist view.* London: Academic.

Mistler-Jackson, M., & Songer, N. B. (2000). Student motivation and Internet technology: Are students empowered to learn science? *Journal of Research in Science Teaching, 37*(5), 459–479.

Miyazaki, I. (1976). *China's examination hell.* New York: Weatherhill.

Moje, E. B., Collazo, T., Carrillo, R., & Marx, R. W. (2001). "Maestro, what is 'quality'?": Language, literacy, and discourse in project-based science. *Journal of Research in Science Teaching, 38*(4), 469–498.

Mokros, J., & Tinker, R. F. (1987). The impact of microcomputer-based labs on children's ability to interpret graphs. *Journal of Research in Science Teaching, 24*(4), 369–383.

Morgan, D. A. (1971, November). STEPS: A science test for evaluation of process skills. *The Science Teacher,* 77–79.

Morgenstern, C. F., & Renner, J. W. (1984). Measuring thinking with standardized tests. *Journal of Research in Science Teaching, 21*(6), 639–648.

Morris, N. (1969). An historian's view of examinations. In S. Wiseman (Ed.), *Examinations and English education.* Manchester, UK: Manchester University Press.

Muir, M. (1970). How children take responsibility for their own learning. In V. Rogers (Ed.), *Teaching in the British primary school.* New York: Macmillan.

Mullis, I. V. S., Dossey, J. A., Campbell, J. R., Gentile, C. A., O'Sullivan, C. O., & Latham, A. S. (1994). *NEAP 1992, trends in academic progress: Achievement of U.S. students in science, mathematics, 1973 to 1992; reading, 1971 to 1992; writing, 1984 to 1992.* Washington, DC: Office of Education Research and Improvement.

Mullis, I. V. S., Martin, M. O., Beaton, A. E., Gonzalez, E. J., Kelly, D. L., & Smith, T. A. (1998). *Mathematics and science achievement in the final years of secondary school: IEA's Third International Mathematics and Science Study.* Chestnut Hill, MA: TIMSS International Study Center.

NASA Jet Propulsion Laboratory. (2015). NASA confirms evidence that liquid water flows on today's Mars. Retrieved on September 2015, from http://www.jpl.nasa.gov/news/news.php?feature=4722

National Commission of Excellence in Education. (1983). *A nation at risk: The imperative for education reform* (Stock Number 065-000-001772). Washington, DC: GPO.

National Commission on Mathematics and Science Teaching. (2000). *Before it's too late.* Washington, DC: DoE.

National Education Goals Panel. (1997). *Mathematics and science achievement for the 21st century.* Washington, DC: National Education Goals Panel.

National Research Council. (1996). *National science education standards.* Washington, DC: National Academy.

National Research Council. (2000). *How people learn science: Brain, mind, experience, and school* (Exp. ed.). Washington, DC: National Academy.

National Research Council. (2012). *A framework for K–12 science education: Practices, crosscutting concepts, and core ideas.* Committee on a Conceptual Framework for New K–12 Science Education Standards. Board on Science Education, Division of Behavioral and Social Sciences and Education. Washington, DC: National Academies.

Nedelsky, L. (1965). *Science teaching and testing.* New York: Harcourt, Brace & World.

Next Generation Science Standards Lead States. (2013). *Next generation science standards: For states, by states, Vols. 1 & 2.* Washington, DC: National Academies.

Noddings, N. (1995). *Philosophy of education.* Boulder, CO: Westview.

Northwest Regional Educational Laboratory (1994). Improving science and mathematics education: A database and catalog of alternative assessments. Second Edition, ED 379151. Portland, Oregon: Author.

Northwest Regional Educational Laboratory (1995). Improving science and mathematics education: A database and catalog of alternative assessments. Third Edition, ED 462252. Portland, Oregon: Author.

Novak, J. D. (1993). Human constructivism: A unification of psychological and epistemological phenomena in meaning making. *International Journal of Personal Construct Psychology, 19*(2), 167–193.

Novak, J. D., Cowin, D. B, & Johansen, G. T. (1993). The use of concept mapping and knowledge vee mapping with junior high school science students. *Science Education, 67*(5), 625–645.

Nussbaum, J., & Novak, J. (1976). An assessment of children's concepts of the earth using structured interviews. *Science Education, 60*(40), 535–550.

Oberholtzer, E. E. (1934). Comments by leaders in the field. In G. M. Whipple, (Ed.), *The activity movement. The third yearbook of the National Society for the Study of Education, Part II.* Bloomington, IL: Public School.

Office of Technology Assessment. (1992). *Testing in American schools: Asking the right questions.* Washington, D.C.: U.S. Congress.

Osborne, J., Leach, J., & Scott, P. (1997). Professor Rosalind H. Driver (1941–1997). *Studies in Science Education, 30*(1), 1–4.

Osborne, R., & Freyberg, P. (1992). Constructing a survey of "alternative" views. In R. Osborne, & P. Freyberg (Eds.), *Learning in science: The implications of children's science* (pp. 166–167). Auckland, New Zealand: Heinemann.

Oxford Dictionaries. (2015). Internet. Retrieved from http://www.oxforddictionaries.com/us/definition/american_english/internet

Palmer, S., & Hall, W. (2011). An evaluation of a project-based learning initiative in engineering education. *European Journal of Engineering Education, 36*(4), 357–365.

Papert, S. (1980). *Mindstorms—Children, computers and powerful ideas*. New York: Basic.

Papert, S. (1993). *The children's machine*: Rethinking school in the age of the computer. New York, NY: Basic Books.

Park Rogers, M. A., Cross, D. I., Gresalfi, M. S., Trauth-Nare, A. E., & Buck, G. A. (2011). First year implementation of a project-based learning approach: The need for addressing teachers' orientations in the era of reform. *International Journal of Science and Mathematics Education*, 9(4), 893–917.

Patton, M. Q. (1987). *How to use qualitative methods in evaluation*. Newbury Park, CA: Sage.

Patton, M. Q. (1990). *Qualitative evaluation and research methods*. Newbury Park, CA: Sage.

Patton, M. Q. (2015). *Qualitative evaluation and research methods* (4th ed.). Newbury Park, CA: Sage.

Paulson, F. L., Paulson, P. R., & Meyer, C. A. (1991, February). What makes a portfolio a portfolio? *Education Leadership*, 48, 60–63.

Pearce, J. M. (2007). Teaching physics using appropriate technology projects. *The Physics Teacher*, 45(3), 164–167. dx.doi.org/10.1119/1.2709675

Piaget, J. (1967). *Six psychological studies*. New York: Vintage.

Piaget, J. (1969). *The psychology of the child* (H. Weaver, Trans.). New York: Basic.

Piaget, J. (1971). The theory of stages in cognitive development. In D. Green, M. Ford, G. Flamer (Eds.), *Measurement and Piaget*. New York: McGraw-Hill.

Pizzini, E. L. (1992, March). Alternative assessment in Iowa. *Science Scope*, 15(6), 57.

Polman, J. L. (2000). *Designing project-based science instruction: Connecting learners through guided inquiry*. New York: Teachers College.

Popham, W. J. (2007). *Classroom assessment: What teachers need to know*. New York: Allyn & Bacon.

Raizen, S. A., & Britton, E. D. (1997). *Bold ventures, Vol.: Case studies of U.S. innovations in science education*. Boston: Kluwer Academic.

Raizen, S. A., & Michelsohn, A. E. (1994). *The future of science in elementary schools*. San Francisco: Jossey-Bass.

Rapp, S. (2008). The quiet skies project: Students collect, analyze and monitor data on radio frequency interference. *The Science Teacher*, 75(7), 62–66.

Raven, J. (1992). A model of competence, motivation and behavior, and a paradigm for assessment. In A. R. Torn (Ed.), *Towards new science of educational testing and assessment*. Albany: State University of New York Press.

Redish, E. F., Saul, J. M., & Steinberg, R. N. (2015). *On the effectiveness of active-engagement microcomputer-based laboratories*. Retrieved from https://physics.ucf.edu/~saul/articles/mbl.PDF

Regassa, L. B., & Morrison-Shetlar, A. I. (2009). Student learning in a project-based molecular biology course. *Journal of College Science Teaching*, 38(6), 58–67.

Reiser, B. J., Krajcik, J., Moje, E., & Marx, R. (2003). *Design strategies for developing science instructional materials*. Paper presented at the Annual Meeting of the National Association of Research in Science Teaching, Philadelphia, PA.

Rittenburg, R., Miller, B. G., Rust, C., Esler, J., Kreider, R., Boylan, R., & Squires, A. (2015). The community connection: Engaging students and community partners in project-based science. *The Science Teacher*, 82(1), 47–52.

The riverbank is their classroom: Hudson assignment benefits turtles. *The Boston Globe.* (2007). Retrieved on September 1, 2015, from http://www.boston.com/news/local/articles/2007/06/17/the_riverbank_is_their_classroom.

Rivet, A. E., & Krajcik, J. S. (2004). Achieving standards in urban systemic reform: An example of a sixth grade project-based science curriculum. *Journal of Research in Science Teaching, 41*(7), 669–692.

Rivet, A. E., & Krajcik, J. S. (2008). Contextualizing instruction: Leveraging students' prior knowledge and experiences to foster understanding of middle school science. *Journal of Research in Science Teaching, 45*(1), 79–100.

Roblyer, M. D., Edwards, J., & Havriluk, M. A. (1997). *Integrating educational technology into teaching.* Upper Saddle River, NJ: Merrill.

Roseberry, A. S., Warren, B., & Conant, F. R. (1992). *Appropriating scientific discourse: Findings from language minority classrooms.* Cambridge, MA: TERC Working Paper, 1–92.

Rowntree, D. (1987). *Assessing students: How shall we know them?* New York: Nichols.

Sadeh, I., & Zion, M. (2012). Which type of inquiry project do high school biology students prefer: Open or guided? *Research in Science Education, 42*(5), 831–848.

Sadler, T. D., Burgin, S., McKinney, L., & Ponjuan, L. (2010). Learning science through research apprenticeships: A critical review of the literature. *Journal of Research in Science Teaching, 47*(3), 235–256.

Schinske, J. N., Clayman, K., Busch, A. K., & Tanner, K. D. (2008). Teaching the anatomy of a scientific journal article. *The Science Teacher, 75*(7), 49–56.

Schmidt, W. H., McKnight, C. C., & Raizen, S. A. (1997). *Splintered vision: An investigation of U.S. mathematics and science education.* Norwell, MA: Kluwer.

Schneider, R. M., Krajcik, J., & Blumenfeld, P. (2005). Enacting reform-based science materials: The range of teacher enactments in reform classrooms. *Journal of Research in Science Teaching, 42*(3), 283–312.

Schneider, R. M., Krajcik, J., Marx, R. W., & Soloway, E. (2002). Performance of students in project-based science classrooms on a national measure of science achievement. *Journal of Research in Science Teaching, 39*(5), 410–422.

Scott-Little, C., Hamann, M. S., & Jurs, S. G. (2002). Evaluation of after-school programs: A meta-evaluation of methodologies and narrative synthesis of findings. *American Journal of Evaluation, 23*(4), 387–419.

Seal, K. (1993). Performance-based tests. *Omni,* 16(3), pages unknown.

Shavelson, R. J., & Baxter, G. P. (1992, May). What we've learned about assessing hands-on science. *Educational Leadership, 49,* 20–25.

Shavelson, R. J., Baxter, G. P., & Pine, J. (1991). Performance assessment in science. *Applied Measurement in Education, 4*(4), 347–362.

Shavelson, R. J., Baxter, G. P., & Pine, J. (1992). Performance assessments: Political rhetoric and measurement reality. *Educational Researcher, 4*(4), 347–362.

Sheikh, M., Fulbright, M., & Hademenos, G. (2011). Captain R. Rubber Ducky: A STEM-driven project in aquatic robotics. *The Physics Teacher, 49*(9), 557–559. dx.doi.org/10.1119/1.3661101

Shelly, G. B., Cashman, T. J., Waggoner, G. A., & Waggoner, W. C. (1997). *Discovering computers: A link to the future*. Cambridge, MA: International Thomson.

Shen, Z., Jensen, W., Wentz, T., & Fischer, B. (2012). Teaching sustainable design using BIM and project-based energy simulations. *Educational Sciences, 2*(3), 136–149.

Short, H., Lundsgaard, M. F. V., & Krajcik, J. (2008). How do geckos stick? Using phenomena to frame project-based science in chemistry classes. *The Science Teacher, 75*(8), 38–43.

Sola, A. O., & Ojo, O. E. (2007). Effects of project, inquiry and lecture-demonstration teaching methods on senior secondary students' achievement in separation of mixtures practical test. *Educational Research and Review, 2*(6), 124–132.

Spring, J. (1999). American education, ninth edition. Boston, McGraw-Hill College.

Spring, J. (2001). *The American school: 1642–2000*. Boston: McGraw-Hill.

Stevenson, J. A. (1928). *The project method of teaching*. New York: Macmillan.

Stiggins, R. J. (1994). *Student-centered classroom assessment*. New York: Macmillan.

Stiggins, R. (1995). *Sound performance assessments in the guidance context*. ERIC Digest (073). ERIC Document Reproduction Service No ED388889.

Stiggins, R. (2007). Educating the whole child: Assessment through the student's eyes. *Education Leadership, 64*(8), 22–26.

Stimson, R. W. (1914). *The Massachusetts home project plan of vocational education*. U.S. Bureau of Education, Bulletin 8, Whole Number 579.

Stockton, J. L. (1920). *Project work in education*. Boston: Houghton Mifflin.

Tal, R., & Argaman, S. (2005). Characteristics and difficulties of teachers who mentor environmental inquiry projects. *Research in Science Education, 35*(4), 363–394.

Tal, R., Krajcik, J. S., & Blumenfeld, P. C. (2006). Urban schools' teachers enacting project-based science. *Journal of Research in Science Teaching, 43*(7), 722–745.

Tamir, P. (1974). An inquiry oriented laboratory examination. *Journal of Educational Measurement, 11*(1), 25–33.

Technology and Cognition Group at Vanderbilt. (1990). Anchored instruction and its relationship to situated cognition. *Education Researcher, 19*(6), 2–10.

Tetenbaum, Z. (1992, March). An ordered approach. *Science Scope, 15*(6), 12–18.

Theberge, C. L., Morrison, D. & Crowder, E. M. (1993). Changing how we measure change: Assessing students' science talk (Unpublished Draft). Cambridge, MA: Bolt, Beranek and Newman.

Thomas, J. W. (2000). A review of research on project-based learning. Retrieved from http://www.bobpearlman.org/BestPractices/PBL_Research.pdf.

Thornton, R. K., & Sokoloff, D. R. (1990). Learning motion concepts using real-time microcomputer-based laboratory tools. *American Journal of Physics, 58*(9), 858–867.

Tinker, R. F. (1986). Modeling and MBL: Software tools for science. Cambridge, MA: Technical Education Research Center, Inc.

Tinker, R. F. (1991). Science for kids: The promise of technology. In K. Sheingold & L. G. Roberts (Ed.), *This year in school science, technology for teaching and learning*. Washington, DC: American Association for the Advancement of Science.

Tinker, R. F. (1992). *Thinking about science*. Princeton, NJ: College Entrance Examination Board.

Tinker R. F. (Ed.). (1996). *Microcomputer-based labs: Educational research and standards*. NATO ASI Series F, Volume 156. Berlin: Springer.

Tinker, R. F., & Papert, S. (1989). Tools for science education. In J. D. Ellis (Ed.), *1988 AETS yearbook: Information technology and science education*. Washington, DC: Office of Educational Research.

Tobin, K. (1993). *The practice of constructivism in science education*. Hillsdale, NJ: Lawrence Erlbaum.

Toolin, R., & Watson, A. (2010, April/May). Students for sustainable energy: Inspiring students to tackle energy projects in their school and community. *The Science Teacher*, 27–31.

Trauth-Nare, A., & Buck, G. (2011). Using formative assessment in problem- and project-based learning. *The Science Teacher*, 78(1), 34–39.

Tsaparlis, G., & Gorezi, M. (2007). Addition of a project-based component to a conventional expository physical chemistry laboratory. *Journal of Chemical Education*, 84(4), 668–670.

Tuckman, B. W. (1988). *Conducting educational research* (3rd ed.). New York: Harcourt Brace Jovanovich.

Turner, G. L'E. (1998). *Scientific instruments 1500–1900*. Berkeley, CA: Philip Wilson.

Tyler, R. W. (1981b). Specific approaches to curriculum development. In H. A. Giroux, A. N. Penna, & W. F. Pinar (Eds.), *Curriculum and instruction: Alternatives in education* (pp. 17–30). Berkeley, CA: McCutchan.

USGS (2016). *Global geographical information system: What is a GIS?*. Retrieved from http://webgis.wr.usgs.gov/globalgis/tutorials/what_is_gis.htm

USGS. (2000). *Herpetology Project*. Retrieved on September 5, 2015, from http://fl.biology.usgs.gov/amphibians2.pdf

Vacchina, P., & Aguirre, M. (2008). The herpetology project: Students construct traps to collect and analyze turtle data. *The Science Teacher*, 78(8), 50–55.

Victor, E. & Kellough, R. D. (1997). *Science for the elementary and middle school*. Upper Saddle River, NJ: Merrill.

Vieyra, R., Vieyra, C., Jeanjacquot, P., Marti, A., & Monteiro, M. (2015). Turning your smartphone into a science laboratory: Five challenges that use mobile devices to collect and analyze data in physics. *The Science Teacher*, 82(9), 32–40.

Vygotsky, L. S. (1978). *Mind in society: The development of higher psychological processes*. (M. Cole, V. John-Steiner, S. Scribner, & E. Souberman, Eds.). Cambridge, MA: Harvard University Press.

University of the State of New York. (1996). *Learning standards for mathematics, science, and technology*. Albany: New York State Education Department.

Unpublished Teacher Report (1994). Indoor air quality investigation report. San Antonio, TX: E. M. Pease Middle School.

Wadsworth, B. J. (1989). *Piaget's theory of cognitive and affective development* (4th ed.). New York: Longman.

Walker, G., Wahl, E., & Rivas, L. M. (2005). *NASA and after-school programs: Connecting to the future*. New York: American Museum of Natural History.

Walton, M., & Archer. A. (2004). The Web and information literacy: Scaffolding the use of web sources in a project-based curriculum. *British Journal of Educational Technology*, 35(2), 1–14.

Warren, G. (1979). Essay versus multiple-choice tests. *Journal of Research in Science Teaching, 16*(6), 563–567.

Weizman, A., Schwartz, Y., & Fortus, D. (2008). The driving question board: A visual organizer for project-based science. *The Science Teacher, 75*(8), 33–37.

Whipple, G. M. (Ed.). (1934). *The activity movement.* The third yearbook of the National Society for the Study of Education, Part II. Bloomington, IL: Public School.

Wiersma, W., & Jurs, S. G. (1990). *Educational measurement and testing* (2nd ed.). Boston: Allyn & Bacon.

Wiggins, G. P. (1993). *Assessing student performance: Exploring the purpose and limits of testing.* San Francisco: Jossey-Bass.

Wilhelm, J., & Confrey, J. (2005). Designing project-enhanced environments: Students investigate waves and sound. *The Science Teacher, 72*(9), 42–45.

Wilhelm, J., Thacker, B., & Wilhelm, R., (2007). Creating constructivist physics for introductory university classes. *Electronic Journal of Science Education, 11*(2), 19–37.

Wilson, C. C., Parkin, A., & Thomas, L. H. (2011). Frontiers of crystallography: A project-based research-led learning exercise. *Journal of Chemical Education, 89*(1), 34–37.

Wilson, E. B., Jr. (1990). *An introduction to scientific research.* New York: Dover.

Wolf, D. P. (1988, December–January). Opening up assessment. *Educational Leadership, 45,* 24–29.

Wolf, D. P. (1989). Portfolio assessment: Sampling student work. *Educational Leadership, 46,* 35–39.

Wolf, D. P., LeMahieu, P. G., & Eresh, J. (1992, May). Good measure: Assessment as a tool for educational reform. *Educational Leadership, 49,* 8–13.

Wolk, S. (1994). Project-based learning: Pursuits with a purpose. *Educational Leadership, 25*(3), 42–45.

Wright, R., & Boggs, J. (2002). Learning cell biology as a team: A project-based approach to upper-division cell biology. *Cell Biology Education, 1*(4), 145–153.

Wu, H. K., & Krajcik, J. (2006). Exploring middle school students' use of inscriptions in project-based science classrooms. *Science Education, 90*(5), 852–873.

Zeiler, M. (1999). *Modeling our world. The ESRI guide to geodatabase design.* Redlands, CA: ESRI.

INDEX

A

A Nation at Risk, 13, 19–20
AAAS, 21, 173
Active participation, 16, 95
Activities, 2–4, 7–8, 14, 17, 20–21, 27, 33, 37–38, 42, 47, 51, 65, 68–70, 78, 81–82, 87, 89–90, 93–97, 99–100, 102, 105, 109–110, 121, 135, 141, 147, 149, 156, 167–168
Afterschool, 49, 73, 81, 92, 145–152; *see also* Afterschool PBSI; Afterschool science
Afterschool PBSI, 145–146, 148–152
Afterschool science, 81, 145–148, 150–152
Aguirre, M., 34, 49–50, 90, 94–98
Air quality, 98–102, 110, 165
American Association for the Advancement of Science, 12–13, 19, 21, 173
American Educators, 6
American Library Association (ALA), 154
American Republic, 3
Analyzing project data, 88–89
Applications, 7, 28, 44, 46, 48, 51–52, 54, 110, 157, 159, 161, 163, 167
Applications of PBSI, 28, 44, 51–52
ARA, 28–31, 34, 36, 40, 46, 48, 51
Artifacts, 17–18, 30, 32, 35, 45, 48–49, 54, 70, 90, 116, 118, 156, 164
Assessment, 13, 20–21, 26, 30–34, 36–38, 44–46, 50–51, 54–55, 57, 65, 70, 75–76, 81, 93–95, 97, 105–106, 109–110, 113–118, 121–143, 154–155, 157, 160, 168
Associations, 153–154, 173–174

B

Ballooning, 29, 107–108, 110, 166
Baxter, G. P., 116, 121, 136–138, 140–142
Beck-Winchatz, B., 29, 46, 108–110
Benefits, 32, 43, 56, 96, 107, 110, 142, 145, 150–151, 153, 170

Biology, 19, 26–27, 30–34, 38, 44–46, 65, 93–94, 97, 108, 129, 133, 165, 172–173
Black, P., 4, 85, 96, 115, 117, 136
Blumenfeld, P., 28, 36–37, 41–42, 76, 116
Bowles, S., 3–5
Bramble, J., 108–109
Brown v. Board of Education of Topeka, 3, 5

C

California Golden State Examination, 125, 139
Carbon dioxide, 98–101, 103, 110
Case Studies, 39, 155
Categorical data, 88
Categories of projects, 18–19
Center for Studies in Science and Mathematics Education, 9
Centers, 153, 156, 174
Challenges, 20, 32, 41–42, 45, 48, 50, 52, 54, 56, 59–60, 69, 76, 81, 84, 90, 93, 98, 102, 106–107, 110, 113, 143, 145–146, 150, 152, 161, 164, 166–167
Characteristics, 22, 37, 48, 57, 85–87, 103, 131, 145–146, 152
Chase, S. E., 2, 15–16, 58
Chemistry, 19, 26–27, 29–31, 34–36, 38, 40, 44, 46–48, 83, 93–94, 97–98, 100, 108, 164, 172
Child-centered, 5–8, 12, 22
Children's Learning in Science Project, 9
CODAP, 162–163
Collaboration, 14, 17–18, 56, 59, 75, 77, 80, 94, 97, 99–100, 104, 129, 149, 156, 168
Community, 4, 10, 14–15, 17, 22, 30–31, 39, 43, 48, 54, 56, 74, 79, 93–99, 107, 122, 135, 142, 145–152, 163, 167–168, 170, 174
Community engagement, 145, 148–149
Computer hardware, 153, 157–158, 174
Computer software, 82, 84, 97, 99, 157

Concepts, 10, 17–18, 21–22, 46, 48–49, 51, 53–54, 62, 65, 67, 69, 83–84, 103, 106, 125, 139, 143, 146, 150, 157, 162
Concepts of PBSI, 53–54, 69
Connecticut Academic Performance Test, 139
Constructivist, 8–11, 14, 35, 99, 136
Contextual factors, 20, 23, 43, 73, 85
Continuous data, 88
Curricula, 13–15, 28, 32, 41, 68–69, 76, 82, 84, 113, 118, 121, 123, 127–129, 133–134, 136, 140, 143, 147, 169–171
Curriculum materials, 21, 28, 51–52, 82, 153, 160, 163, 169–170, 174

D

Dartmouth College, 3
Data, 14–15, 18, 27, 34, 39–41, 43–46, 49, 59, 61, 63, 70, 73, 81–89, 91–92, 95–98, 100–106, 108–110, 119–120, 129, 141, 149, 151, 155, 157–163, 166, 169, 171–172
Data analysis, 14, 40, 49, 59, 63, 70, 88–89, 91, 95–96, 100, 106, 160, 163
Data collection, 40, 43, 59, 63, 70, 81, 84–85, 89, 91–92, 95–96, 100–101, 104–106, 108–110, 149, 159
Database, 14, 25, 94, 102, 113–114, 153–154, 158–159, 165, 171, 174–175
Decision tree, 88–89, 92
Definitions, 1, 15–18, 115, 131, 155, 164
Dependent variable, 62–63, 67, 126
Dewey, J., 7–8, 12
Discussion, 1, 3, 38, 43, 45–46, 59, 63, 65, 73, 90–91, 109, 115, 121, 127, 131, 143, 145, 149, 152, 159, 164
Dispositions, 22, 42, 45, 47, 53, 55, 62, 67, 105–106, 110, 146, 150, 154
Drawings, 29, 88, 114, 117–118, 129–130
Driver, R., 3, 9–11

INDEX 197

Driving question, 17, 34, 55, 61, 65, 68, 80, 90, 105, 108
Driving research question, 48–49, 63, 67–68, 85, 104

E

Earth Science, 26, 49, 65, 81, 85, 171
Ecology, 36, 49, 103–106, 110
Economic, 3–5, 113, 123, 127–128, 134–135, 139, 141–143
Education, 2–14, 16–17, 19–23, 26, 31, 33, 38, 40, 50–53, 76–77, 82–83, 94–95, 114, 116, 124–126, 129, 131–132, 136, 139, 142, 145–148, 150–154, 156–157, 160–163, 165–174
Empirical research articles, 26
Engagement, 2, 4, 6, 8, 10, 12, 14, 16, 18, 20, 22, 26, 28, 30, 32, 34, 36, 38, 40–46, 48, 50–52, 54, 56, 58, 60–62, 64, 66, 68, 70, 74, 76, 78, 80, 82, 84, 86, 88, 90, 92, 94, 96, 98, 100, 102, 104, 106, 108, 110, 114, 116, 118, 120, 122, 124, 126, 128, 130, 132, 134, 136, 138, 140, 142, 145–146, 148–150, 152, 154, 156, 158, 160, 162, 164, 166, 168, 170, 172, 174
Engagement in PBSI, 42–43
Engagement of teachers, 28, 41, 51–52
Engineering, 21–22, 26, 29, 31, 37–38, 44, 50–51, 54, 107–108, 148, 154, 164, 169
Environmental Science, 26–27, 44, 49, 85, 93, 97, 103, 105–106
Environmental Systems Research Institute, 163
ERA, 4–5, 27–40, 46, 48, 50–51
Essay, 28–34, 37–40, 57, 65, 114–118, 124–128, 130, 135, 138
Evaluation, 12, 26–27, 30–31, 34–35, 38, 40, 45, 49, 68, 98–99, 104, 109, 113–115, 132, 134, 137, 143, 146–147, 168
EVRA, 27–31, 33, 35, 38–40, 46, 48, 50–51
Expectations, 13, 19, 53, 55–57, 77, 103, 132, 142, 149, 151, 161
Expectations of PBSI, 56, 103

F

Facilitators, 41–42, 54, 69, 79, 104, 109
Factors, 12, 20, 23, 29, 41–43, 49, 63, 73–74, 81, 85, 87, 92, 122, 126, 136, 146, 148, 150, 157
Framework for K-12 Science Education, 21–22
French model, 3
Froebel, F., 5–6

G

Geographical Information System (GIS), 163
Gintis, H., 3–5
Global Lab Project, 15, 49, 98, 169
GLOBE Program, 170
Google, 25, 60, 91, 114, 155, 158
Government, 110, 139, 147–148, 153, 155, 170, 172, 174
Group discussion, 46, 109, 121
Groups, 8, 13, 29, 31, 36, 38, 44, 50–52, 59, 75–76, 79–80, 82, 84, 88, 90, 97, 101, 104, 106, 108, 118, 134–135, 143, 151, 154, 156; *see also* Group discussion
Guest speakers, 174
Gwynne-Thomas, E. H., 2–3

H

HAB, 108–110
Hardware, 76, 82–84, 95, 97, 103–104, 106, 108, 153, 157–158, 174
Herbart, J. F., 5–6
Herpetology Project, 34, 93–95, 110
High-altitude ballooning, 107–108, 110, 166; *see also* HAB
Hike, N., 29, 46, 108–110
Historical foundations, 2; *see also* History
History, 1–3, 14, 16, 19, 84, 93, 95, 97, 99, 101, 103, 105, 107, 109, 111, 136, 156, 164

Holistic scoring, 125, 127, 138–139
Hosic, J. F., 2, 15–16, 58
Human Constructivist, 10–11
Hypothesis, 27, 46, 57, 61–63, 91, 96, 99, 119, 123, 129, 159, 163

I

Impact, 14, 27–29, 32, 43, 51–52, 95, 98, 133, 146, 148, 162
Implementation plan, 58, 99, 147
Implementation strategies, 49, 73, 93, 110
Independent variable, 62–63, 67
Indoor air quality, 98, 100, 110
Industrial Revolution, 2
Informal science, 145, 147, 153, 174
Inquiry-based science, 11, 20–21, 37, 64, 114–116
Inquiry-based science environments, 11, 114, 116
Inquiry-based science learning, 20, 115
Instructional materials, 15, 98–100, 104, 165, 168–169
Instrument, 44, 47, 54, 59, 63, 70, 77, 81, 85, 91, 95, 102, 104, 115, 129, 141, 164
Internet, 14, 60, 77, 84, 91, 106, 147–148, 153, 155–156, 158, 171, 174

J

James, W., 7, 168

K

K-12 Science Education, 21–22, 167
K-16 STEM Education, 169, 172
Kids Network Project, 15, 169
Kilpatrick, W. H., 2, 8, 12, 15–16, 18
Knowledge, 3, 7, 9–15, 17–18, 22–23, 28, 32–33, 36, 41–42, 45–46, 48–50, 53, 55, 57, 59–60, 62, 64–66, 74, 78, 83, 89, 94, 98, 105, 110, 118–119, 121, 127, 130, 133, 135, 140, 143, 147–148, 157, 161
Krajcik, J., 16–17, 26, 28–29, 32, 34–39, 41–43, 46, 75–76, 78, 83, 116

L

Lab, 15, 20, 30, 35, 40, 46–49, 77, 82–83, 98, 100, 115, 154, 165, 169
Learning activities, 110
Learning environments, 21, 113–115, 163
Learning science, 9–11, 57, 99, 123, 150
Learning standards, 1, 13, 19–21, 23, 28, 62, 64–65, 67–68, 79, 81, 101, 108, 168
Lecture, 20, 35, 54, 109, 154
LEGO, 167
Lesson, 6, 27, 29, 36, 39, 54, 93, 98, 102, 106, 110, 132, 134–135, 154
Lesson learned, 93, 98, 102, 106, 110
Lesson plans, 6, 154; *see also* Lesson
Library, 60, 78, 153–156, 172, 174
Life of a project, 55
LOGO, 14, 161

M

Madaus, G. F., 115–116, 124
Marx, R., 16, 28, 37–39, 41–43, 75–76, 116
Materials, 2, 9, 15, 21, 28, 35, 37, 49, 51–52, 54, 58, 60, 63, 65–66, 68, 76–78, 81–84, 98–100, 103–105, 126, 129, 132, 134–135, 137–138, 141, 147, 151, 153, 155, 159–161, 163, 165, 167–170, 174
Mentors, 42, 55, 69, 79, 104, 109
Microcomputer-Based Laboratories (MBL), 14, 82–84, 92, 165
Microcomputer-based laboratory tools, 85; *see also* Microcomputer-Based Laboratories (MBL)
Mintzes, J. J., 10–11

Misconceptions, 69, 114, 121, 130, 139, 168
Modeling, 14, 44, 50, 82, 94, 160–164, 175
Models, 14, 22, 36, 47, 122, 125, 127, 132, 159, 162, 165; *see also* Modeling
Moje, E., 16–17, 38, 43, 76
Multiple teachers-multiple classrooms, 73, 80
Museums, 145, 156

N

National Aeronautics and Space Administration (NASA), 107, 110, 147, 167, 170, 172
National Commission of Excellence in Education, 12–13
National Research Council, 11, 13, 19, 21, 64, 172
National Science Digital Library, 154
National Science Education Standards, 13, 19, 94, 154
National Science Foundation, 13, 22, 98, 136, 172
National Science Teachers Association, 13, 17, 21, 138–139, 173
Next Generation Science Standards (NGSS), 1, 21–22, 49, 64, 71, 108, 149, 154, 162–163
Novak, J. D., 10–11, 118
NSDL, 154–155
NSES, 19–22, 64, 71
NSF, 13–15, 41, 169–172

O

Objective-driven, 12
Observations, 29, 33, 36–38, 40–41, 44, 50, 66, 82, 96, 116–117, 170
One teacher-one classroom, 73, 80
Online resources, 153, 155–156, 174
Oral presentation, 108, 118, 121–123, 127–128, 130, 134
Organizations, 5, 13, 23, 148, 151, 153, 156, 169, 171, 174
Orientation to PBSI, 31, 42, 53, 55–57, 69–70, 105
Ownership, 22, 46–47, 68

P

Papert, S., 11, 14, 77, 161
Parental involvement, 74, 77, 149–150; *see also* Parents
Parents, 2, 48, 69, 102, 104, 106, 131, 134, 142, 148, 150–151, 154, 161
Park, 30–31, 41–42, 45, 49, 61, 74, 78, 103–106, 110, 147, 156, 172
Park ecology, 49, 103–104, 106, 110
Participation, 4, 16, 44, 46, 95, 131, 150, 173
Pasco Scientific, 84, 165–166
PBSI, 1, 10–11, 14, 21–23, 25–39, 41–58, 64, 68–71, 73–74, 76–77, 79–81, 84–85, 88–90, 92, 102–104, 107, 110–111, 114–115, 143, 145–154, 156–165, 167–172, 174–175
PBSI curriculum, 28, 41, 43, 49, 51–52, 76, 169–171
PBSI environment, 10, 26, 44, 68–69, 71, 80, 84, 88–89, 111, 143, 147, 158–159, 165
PBSI professional development, 28, 41, 51–52
PBSI research, 26–27, 51–52
Peirce, C., 7
Performance, 13, 31, 38, 47, 55, 57, 95, 113–117, 127, 130–131, 133, 135–143, 162, 168
Performance tasks, 136–140, 142
Performance-based assessments, 13, 168
Pestalozzi, J., 5–6
Philosophical foundations, 5
Physics, 19, 26, 30–31, 35–36, 39, 44, 48–49, 83, 93, 107–108, 148–149, 164, 166, 173
Piaget, J., 8–9

Pine, J., 88, 116, 121, 136–138, 140
Plans, 6, 55, 58, 65, 67–68, 75, 79, 154
Polman, J. L., 12, 78
Portfolios, 36, 57, 114, 116, 118, 130–131, 133–135
Potential benefits, 107, 110, 142, 145, 150, 153
Pragmatists, 7
Principal, 73–74, 79–80, 101, 120
Principles, 4, 6, 10, 16, 22, 47, 53–54, 65, 69, 83, 125, 164
Probes, 82, 99; see also Probeware
Probeware, 82, 84–85, 158
Professional development, 20–21, 26, 28, 34, 36, 41, 51–52, 76, 81, 89, 160, 167, 169–170, 173–174
Professional science associations, 173
Progressive education, 7–8, 22
Project 2061, 19–20, 35
Project-based approach, 2, 5, 8, 15, 47, 49–50, 169
Project-based biology, 93
Project-based chemistry, 38, 40, 46–48, 98
Project-based earth/environmental science, 103
Project-based physics, 48–49, 107
Project-based science instruction, 1, 3, 5, 7, 9, 11–19, 21–23, 25–26, 36, 53, 55, 57–59, 61, 63, 65, 67, 69, 71, 73–81, 91, 93, 95, 97–99, 101, 103, 105, 107, 109, 111, 113, 115, 117, 119, 121, 123, 125, 127, 129, 131, 133, 135, 137, 139, 141, 143, 145, 147, 149, 151, 153, 155, 157, 159, 161, 163, 165, 167, 169, 171, 173, 175
Project-based science learning, 14–15, 26, 28, 51–52, 113, 115, 143
Projects, 2, 8, 11–15, 17–19, 22–23, 29, 33, 35–37, 42, 45, 47–51, 53, 55–59, 63, 67–70, 74–82, 84, 89–90, 97, 99, 102–103, 105–110, 114, 129, 147–148, 150, 156, 158–159, 163, 168–170
Project activities, 69–70, 90, 93, 97, 102, 105, 109
Project cycle, 42, 53, 56, 69, 71, 146, 151, 154, 159, 168–169
Project data, 84–85, 88–89, 92, 161–162; see also Data
Project plans, 55, 65, 67–68, 79
Project questions, 60–62, 64–67, 84
Project reports, 49, 65, 89–90, 92, 106, 114, 118, 128–129
Psychological foundations, 1, 8
Purposeful act, 15

R

Relationships, 11, 78, 124, 129, 145, 147, 151, 159, 163
Relevance, 8, 10, 28, 51, 102, 148
Reliability, 121, 128, 140
Requirements, 55, 57, 64, 67, 79, 82, 163
Research, 8–9, 11, 13–15, 18–19, 21–23, 25–29, 31, 33, 35, 37, 39–41, 43, 45, 47–52, 57, 60–61, 63–64, 66–68, 81–82, 85, 88–91, 94–96, 98–100, 102–106, 108–109, 114, 119–120, 129, 136, 138, 140, 142, 147, 149, 151, 153–154, 156, 159, 163, 165, 169–174
Research centers, 153, 174
Research process, 106, 119
Research question, 18, 22–23, 27, 48–49, 52, 61, 63, 67–68, 85, 90–91, 103–105, 109, 119, 129, 154, 156
Responsibilities, 47, 53, 55, 58, 66, 71, 75, 89, 94, 105, 147, 154
Robotics, 31, 107, 167–168
Role of teacher, 109
Roles, 21, 45, 50, 53–55, 58, 66, 71, 74, 79, 93–95, 97, 101, 105, 110–111
Rousseau, J. J., 5–6
Rubrics, 168

S

Schneider, R. M., 28, 37–38, 43–44, 76, 161
Science content, 14, 20–21, 41, 47, 49, 54–55, 64, 81, 97, 100, 119, 123, 127, 146–147

Science disciplines, 28, 44, 51–52, 125, 154
Science education associations, 153, 173–174
Science education research, 23, 26
Science education standards, 13, 19, 94, 154
Science learning, 2, 4, 6, 8, 10, 12, 14–16, 18–23, 26, 28, 30, 32, 34, 36, 38, 40–44, 46–48, 50–52, 54, 56, 58, 60, 62, 64–68, 74, 76, 78–80, 82, 84, 86, 88, 90, 92, 94, 96, 98–100, 102, 104, 106, 108, 110, 113–116, 118, 120, 122–124, 126, 128, 130–132, 134, 136, 138, 140, 142–143, 146, 148, 150–152, 154, 156–158, 160, 162, 164, 166, 168, 170, 172, 174
Science learning standards, 19, 21, 23, 28, 64–65, 67, 79, 108
Science process skills, 13, 15–16, 21, 36, 76, 80, 83, 106, 114, 119–120, 124, 126, 129–131, 133, 136, 139, 141, 150
Science research, 129, 171
Science unit, 41, 121; *see also* Unit
Scientific electronic databases, 171, 175
Scientific investigations, 43, 59, 104, 119
Scientific research, 15, 81, 119, 149, 153, 159, 171–172, 174
Scientific research centers, 153, 174
Scientific tools, 15, 97–100, 153, 164, 174–175
Scoring procedures, 138, 142
Scoring protocol, 118, 121, 138, 140
Scoring system, 118, 125, 136
Sensors, 82, 85, 165
Shavelson, R. J., 116, 121, 136–138, 140–142
Smith Agricultural School, 11
Social, 2, 4, 7–10, 15–16, 43, 45, 63, 79, 98, 113, 119, 122–123, 125, 127–129, 133–134, 141–143, 148, 155, 160, 163–164
Software, 29, 50, 76–77, 82–84, 90, 95, 97, 99–100, 103–104, 153, 155, 157–163, 165–167, 171, 174–175
Soloway, E., 28, 37–39, 76, 116
Standards, 1–2, 12–13, 19–23, 28, 35, 37, 45–46, 54, 61–62, 64–65, 67–68, 75, 79, 81, 94, 97, 100–101, 108, 122, 124, 148, 154, 168
Statistical analysis, 96, 159–160
Statistics, 126, 159–161
STELLA, 162
STEM, 26, 29, 31, 41, 44, 114, 147, 154, 161, 163, 167–172
Stevenson, J. A., 11, 15–17, 129
Stiggins, R. J., 116, 121–123, 125, 128, 132, 136
Stimson, R. W., 11
Stockton, J. L., 5, 7, 15–16, 129, 136
Strategies, 11, 34, 49, 59–60, 65, 73–74, 79–80, 91–93, 97, 102, 105, 109–110, 121, 135, 143, 151, 154
Student engagement, 42–43
Student projects, 14, 48, 114
Student-centered, 9, 13, 29, 73, 79, 91, 97, 105, 109, 114, 163
Students' lives, 28, 54
Subject matter, 12

T

Tangible outcomes, 18, 23, 54
Teacher's role, 93–94, 101, 103, 109
Teacher-centered, 7, 41, 73, 79–80, 91
Teacher-student-scientist partnership, 80–81, 91
Technical, 21, 47, 50, 97, 113, 121–122, 127–128, 133–134, 140–141, 143, 147, 149, 164, 169
Technology and Cognition Group at Vanderbilt, 11
Testing, 2, 14, 59, 61, 66–67, 70, 79, 81, 86–87, 99–100, 103–106, 114–115, 117, 127–128, 134–136, 140–143, 159, 165
Theory, 5, 11, 14, 28, 31, 37, 40, 45, 48, 50, 65, 88, 119, 121, 125, 142–143, 161
Thomas, J. W., 2–3, 31, 39, 46
Tinker, R. F., 14, 16–17, 22, 77, 82–83
Tools for PBSI, 163
TRA, 28–34, 37, 39–40, 46, 51

U

U.S. Education, 7, 172
Underlying principles, 22, 53–54, 83, 164
Unit, 32–33, 41, 46–47, 49–50, 68–69, 82, 87, 94, 107, 115, 121, 160
United Nations (UN), 171
United States Geological Survey (USGS), 93, 163, 172
University of Georgia, 3
University of Leeds, 9
University of Michigan, 26
University of North Carolina, 3
University of the State of New York, 65

V

Vacchina, P., 34, 49–50, 90, 94–98
Validity, 51, 134
Variable, 26, 61–63, 67, 82–83, 88, 91–92, 119–120, 126, 159, 162–163, 171
Vendors, 84, 108, 162, 165, 174
Vernier Software and Technology Company, 84, 165–166
Vygotsky, L. S., 8–9

W

Wandersee, J. H., 10–11
Weaknesses, 59, 113, 122, 128, 130, 133–134, 141, 143

www.ingramcontent.com/pod-product-compliance
Ingram Content Group UK Ltd.
Pitfield, Milton Keynes, MK11 3LW, UK
UKHW021845140426
5217IPUK00022B/1603